# Isotopes Applications in Earth Sciences

*Edited by Rehab O. Abdel Rahman*

Published in London, United Kingdom

IntechOpen

*Supporting open minds since 2005*

Isotopes Applications in Earth Sciences
http://dx.doi.org/10.5772/intechopen.77560
Edited by Rehab O. Abdel Rahman

Contributors
Vicente Albino Manjate, Minghui Li, Saed Khayat, Amer Marei, Anne Leclerc, Gregory McCarty, Xia Li, Sangchul Lee, Xiaomin Fang, Jiao Li, Maodu Yan, Shurui Sun, Liping Zhu, Zaher Barghouthi



First published in London, United Kingdom, 2020 by IntechOpen
IntechOpen is the global imprint of INTECHOPEN LIMITED, registered in England and Wales, registration number: 11086078, 7th floor, 10 Lower Thames Street, London, EC3R 6AF, United Kingdom
Printed in Croatia

British Library Cataloguing-in-Publication Data
A catalogue record for this book is available from the British Library

Additional hard and PDF copies can be obtained from orders@intechopen.com

Isotopes Applications in Earth Sciences
Edited by Rehab O. Abdel Rahman
p. cm.
Print ISBN 978-1-78985-827-3
Online ISBN 978-1-78985-828-0
eBook (PDF) ISBN 978-1-83881-015-3

For EU product safety concerns:
IN TECH d.o.o., Prolaz Marije Krucifikse Kozulić 3, 51000 Rijeka, Croatia, info@intechopen.com or visit our website at intechopen.com.

# Meet the editor

Rehab O. Abdel Rahman is Associate Professor of Chemical Nuclear Engineering at the Radioactive Waste Management Department, Hot Laboratories and Waste Management Center, Atomic Energy Authority of Egypt. She has a PhD in Nuclear Engineering from Alexandria University, Egypt. She has contributed to the publication of more than thirty peer-reviewed scientific papers, fourteen book chapters and eight books. Dr. Rahman has contributed in teaching and supervising postgraduate research in chemistry, physics, petrochemicals and environmental–chemical engineering. She serves as a verified reviewer for several journals and is managing editor for the *International Journal of Environment and Waste Management* and *International Journal of Engineering Education*. Dr Rahman is an honored scientist of Academy of Scientific Research and Technology and a Publons top reviewer (2016–2019).

# Contents

# Preface

The application of isotopes in exploring different ecological, biological, chemical and geological systems is receiving considerable attention. These applications aim to understand the dynamics of different environmental compartments including hydrosphere, atmosphere, and lithosphere, explore natural resources, plan for water management, assess and predict climatic changes and trace the migration of pollutants. This book provides a wide scope of topics related to the principles and recent advances in the applications of isotopes in the earth sciences. It describes the theoretical background on radioactivity phenomenon, including the identification of the radiological characteristics of natural and anthropogenic isotopes and isotope fractionation processes. It also examines the applications of these isotopes to understand the natural processes in different compartments of the Earth.

The book consists of two sections that cover important research efforts in applying isotopic techniques in earth science. The first section presents some isotopic applications in the lithosphere. Dr. Li et al. present an overview of the formation, sources, transport and fractionation of different natural and anthropogenic isotopes. Their applications in understanding soil redistribution and sedimentation mechanisms are illustrated by demonstrating the feasibility of applying isotopes and icescape modeling. Dr. Manjate uses metamorphic zircon morphology, Th/U ratio, and 207Pb/206Pb to date the tectono-metamorphic event in Macossa-Chimoio Nappe Mozambique.

The second section illustrates some isotopic applications in the hydrosphere. Dr. Li et al. present the applications of stable sulfur, strontium, boron, oxygen and deuterium to study brine origin. Their application is directed to study Qaidam Basin, NE Tibetan Plateau. Dr. Khayat et al. apply stable isotopes to identify recharge–discharge zones and study groundwater flow regimes and the connection between sub-basins. Their applications are directed to support planning for water management by combining isotope analysis with the geological and hydrogeological setting in the West Bank, Palestine. Finally, Dr. Leclerc applies the concept of oxygen isotopic fractionation to investigate the effect of temperature and light intensity on biological and isotopic changes.

I would like to thank all the authors for their efforts that led to the production of this distinguished scientific contribution. A special acknowledgment is directed to Author Service Manager Mr. Mateo Pulko for his coordination efforts.

**Rehab O. Abdel Rahman**
Atomic Energy Authority of Egypt,
Cairo, Egypt

Section 1

# Application in Lithosphere

Chapter 1

# Isoscape Analysis for Elucidating Relationships between Soil Redistribution and Soil Carbon Dynamics

*Xia Li, Gregory McCarty and Sangchul Lee*

## Abstract

Isotopic tracers are useful for assessing linkages between soil movement and soil carbon dynamics in landscapes. Analyses of isotopes and comparison of isoscape (isotopic landscape) with observational data have been employed to investigate spatial distributions of isotopes, to test efficiencies of isotopic models, and to examine soil redistribution patterns and C dynamics. This chapter reviewed the application of natural ($^{7}$Be, $^{210}$Pb) and anthropogenic fallout radionuclides ($^{137}$Cs, $^{239,240}$Pu), and C isotopes ($^{12,13,14}$C) in understanding mechanisms of soil redistribution and sedimentation. The chapter was organized to cover the formation, sources, and transport of these isotopes; how they are distributed and related to soil redistribution on C dynamics; and importance of their distribution (isoscapes) on investigating soil properties. We also provided a case study to demonstrate the feasibility of applying isotopes and isoscape modeling for understanding soil property variability in response to anthropogenic disturbance in a low-relief cropland field. Results demonstrated advantages of using $^{137}$Cs and C isotopic signature ($\delta^{13}$C) to trace soil movements and C dynamics. Topography-based $^{137}$Cs and C isoscape models were developed using light detection and ranging data (LiDAR) derived topographic metrics. The models successfully simulated the spatial patterns of $^{137}$Cs inventory and $\delta^{13}$C over an agricultural landscape and can benefit soil sedimentation and C dynamic studies in areas with limited observations.

**Keywords:** isotope analysis, isoscape, soil redistribution, soil carbon dynamics, dynamic replacement

## 1. Introduction

Soil redistribution (i.e., erosion and deposition) exerts a significant impact on the global carbon (C) cycle. Movement of soil particles could either redistribute soil C or change the C mineralization through disruption of soil aggregation, reaggregation of deposited soils, and deep burial of C-enriched sediments at depositional sites [1]. When soil erosion occurs, the surface concentrated and low-density proportions of sediments, such as soil organic carbon (SOC), are preferentially removed by runoff, wind, and/or tillage activities. A greater SOC enrichment ratio in eroded sediment is usually observed compared to its origin [2]. The eroded sediment is either redistributed over a landscape or deposited at depressional sites. Therefore, investigation of soil redistribution and underlying mechanisms are needed for better understandings of the fate of soil C within the landscape.

Various methods have been developed to quantified soil redistribution. Before the 1990s, researchers mainly focused on understanding soil erosion by discerning changes in soil texture and chemical properties [3]. However, accuracy of this method was low, mainly because impacts of erosion on soil properties are complicated and vary with soil characteristics, soil depth, and local land use practices. Other studies adopted erosion prediction models, such as the USLE, WaTEM, and WEPP, to assess soil redistribution processes [4, 5]. Although these models reasonably estimated long-term mean erosion rates over large-scale, the estimates from the modeling are often mean redistribution rates at sites, model estimates from individual erosion events are too coarse to be linked to each soil samples to explain changes in soil properties [6].

The emergence of isotopic analysis allows researchers to accurately trace soil movement at locations, offering a potential way to quantify the impacts of soil erosion and deposition on soil properties. The isotopic tracer selection follows two criteria: (1) isotopes should be able to quickly and strongly absorbed by the soil; and (2) the variability in absorption to various textures or size is either minor or can be calculated [7]. A tracer can be uniformly distributed at first and then move with the soil movements, presenting mass variations between eroded and depositional sites.

This chapter examines the application of six widely used tracers, including anthropogenic fallout radionuclides/Cesium-137 ($^{137}$Cs), plutonium isotopes ($^{239,240}$Pu), natural fallout radionuclides Beryllium-7 ($^{7}$Be), Lead-210 ($^{210}$Pb), and C isotopes ($^{12,13,14}$C), in soil redistribution and C distribution studies. Objectives of this chapter are to (1) review impacts of soil redistribution on soil C dynamics; (2) summarize critical processes regulating the selected soil isotopes; (3) introduce the concept of isoscape and its applications in soil biogeochemical studies; and (4) provide discussion of isotopic and isoscape analyses for understanding of soil redistribution and carbon dynamics through a case study.

## 2. Soil redistribution impacts on soil C dynamics

The global soil inventory represents an important C pool (**Figure 1**) with a total of 1950 Gt C, accounting for 2.3 times the size of the atmospheric C pool (860 Gt C) and 3.5 times of the biotic C pool (550 Gt C) [8]. About 1750 Gt C of total soil C concentrates on land surface, where soil erosion and deposition processes are intense. The light and fine soil particles with high SOC content are preferentially removed by the erosion process and redistributed over a landscape or deposited at depressional sites. It was estimated that 75 Gt year$^{-1}$ of soil is removed by water and wind erosion [1], which contribute to 0.8–2.2 Gt year$^{-1}$ emission of C from land surface, and 2.5–3.9 Gt year$^{-1}$ of replaced C in soil [8].

Although researchers have increasingly recognized the importance of soil erosion and C dynamics on ecosystems, mechanisms of soil redistribution controls on soil C remains poorly understood. Some studies suggested negative effects of soil erosion on terrestrial C sequestration [9–11]. They provide an argument that excessive soil erosion can lead to losses of soil fertility, decreasing the plant and crop productivity [9]. Notably, the recently accelerated soil erosion due to agricultural activities has caused 2 billion ha of land being irreversibly degraded [1]. Plowing activity increases the possibility of soil organic matter within the plow layer to be exposed to anthropogenic and climatic perturbations, and thus accelerates SOC mineralization [11]. Furthermore, soil C mineralization in displaced soils can also

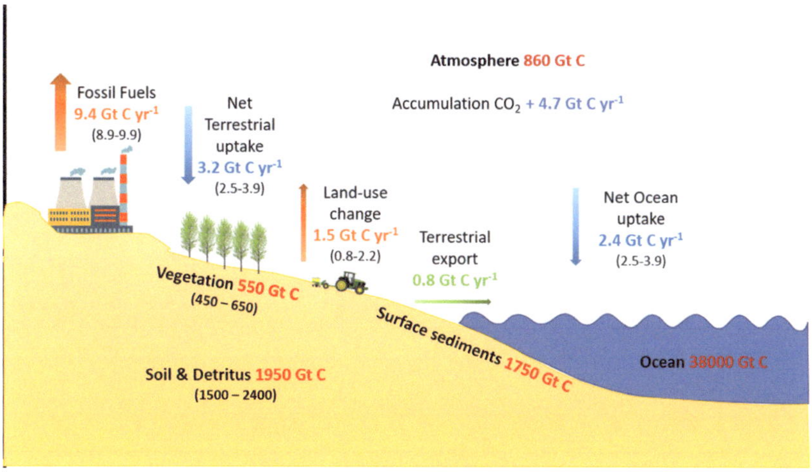


**Figure 1.**
*Global C budget for the decade 2008–2017. The C storage and dynamics were taken from Le Quéré et al. [8].*

be stimulated as a result of the breakdown of soil aggregates by raindrop and runoff shearing forces that expose previously protected organic matters to oxidation [12]. It was estimated that more than 20% C emission was caused by mineralization of the displaced C [13].

On the contrary, arguments for positive impacts of soil redistribution on C storage support that redistribution of soil could increase C sequestration in terrestrial ecosystems [14–17]. SOC losses at the eroded sites could be dynamically replaced by litter input from plant regrowth and return of above- and below-ground biomass to soil and replenish the depleted SOC pool at eroded sites [18, 19]. The eroded SOC is subsequently buried and re-aggregated at depressional sites, protecting SOC from mineralization and thus increasing C sequestration [14, 20, 21]. Stallard [19] used a large set of scenarios estimated that the human-induced C burial is in the range of 0.6–1.5 Gt C year$^{-1}$ by terrestrial sedimentation.

## 3. Using isotopes in studies of soils and sediments

### 3.1 Anthropogenic fallout radionuclides

$^{137}$Cs is the most widely used isotope in geomorphic studies of water, wind and tillage erosion. As an anthropogenic radionuclide with a half-life of 30.2 years, $^{137}$Cs was released globally into the environment due to radioactive fallout from nuclear weapon testing in the 1950s and 1960s and reached to land surface through wet and dry deposition (**Figure 2**). Additional $^{137}$Cs fallout occurred because of the 1986 Chernobyl accident in the north of the Ukrainian SSR and the 2011 Fukushima accident in Japan. The overall amount of $^{137}$Cs released from Fukushima was less than 15% of the amount from the Chernobyl accident [22]. The Fukushima accident added high amounts of $^{137}$Cs to soils near the accident site, but the large-scale plume of radioactive fallout extended mainly over the Pacific Ocean which reduced the extent of terrestrial labeling. In contrast, Chernobyl is in the center of the European continent. The released fallout radionuclides significantly increased the $^{137}$Cs amount in many European countries [22]. $^{137}$Cs is chemically active and is rapidly and strongly absorbed by fine soil particles when it contacts with the soil. Vertical migration is slow and the majority of $^{137}$Cs is retained in the upper 20 cm of the soil surface across the globe [23–25]. Once the soil is labeled by $^{137}$Cs, chemical and

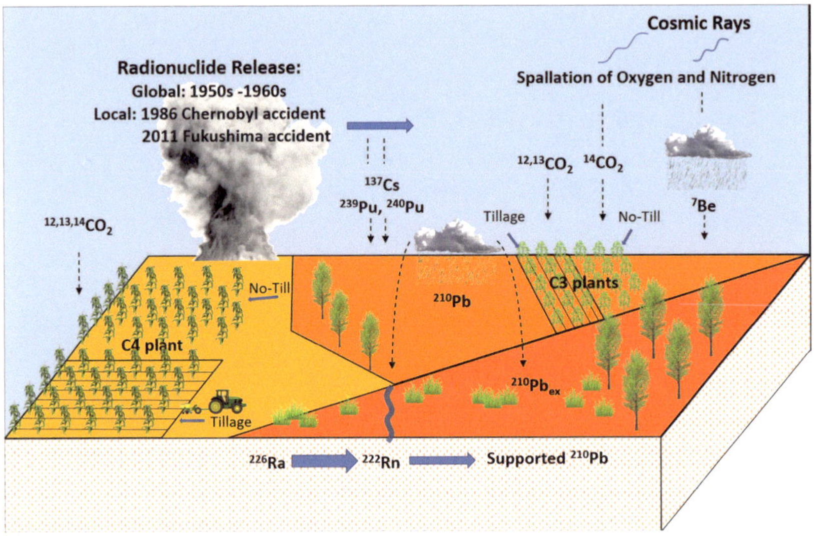


**Figure 2.**
*Isotopes used as soil tracers in soil and sediment studies.*

biological removal of $^{137}Cs$ is low. The $^{137}Cs$ concentration changes mainly result from physical processes in the top-layer soils, such as soil transport and deposition. Therefore, these characteristics make $^{137}Cs$ a useful tracer for quantifying of soil redistribution rates.

Previous investigations have reported strong and statistically significant correlations between $^{137}Cs$ and soil C [10, 26, 27]. However, because $^{137}Cs$ inventory is usually non-linearly correlated with soil redistribution, the correlations between $^{137}Cs$ inventory and soil C content are usually weaker than the correlations between soil redistribution process and SOC [6]. Consequently, instead of directly quantify SOC based on $^{137}Cs$, most recent studies first converted $^{137}Cs$ measurements to soil redistribution rates, and then used the estimates to link to soil C content. Specifically, the application of $^{137}Cs$ measurements in soil redistribution and C dynamics mainly includes to (1) calculate soil redistribution at a point and link to soil properties of a soil sample from the same point [6, 28, 29]; (2) provide mean annual rate of soil redistribution over 60 years and reflect the erosion history of a site [10, 29]; and (3) present spatial patterns of soil erosion and deposition over the landscape [27, 30].

Because of the short half-life, $^{137}Cs$ concentration may drop below the detection limits quickly and limit use of this tracer in SOC investigations. Researchers are finding alternative tracers with longer half-lives for soil erosion investigation. Anthropogenic radionuclides of two major plutonium isotopes ($^{239}Pu$: half-life = 24,110 years and $^{240}Pu$: half-life = 6561 years) are considered as potential alternatives for this purpose. Like $^{137}Cs$, the Pu isotopes in soils are also mainly from nuclear weapon testing between the 1950s and 1960s following the 1986 Chernobyl accident. However, the isotopic composition of Chernobyl Pu ($^{240}Pu/^{239}Pu$ atom ratio: 0.37–0.41) presents higher ratios compared to Pu from nuclear testing ($^{240}Pu/^{239}Pu$ atom ratio = $0.180 \pm 0.014$), which can be used to distinguish Pu isotopes from the two sources [31]. Some researchers have argued that in soils with high organic matter, Pu might be preferentially absorbed by organic fractions with high molecular weight and potentially lead to a deep vertical migration [32, 33].

Considering the long half-life of Pu isotopes, increasing studies have explored the feasibility of using Pu isotopes in investigations of soils and sediments [34–36]. Schimmack et al. [37] found that the Chernobyl Pu could be a better tracer for soil redistribution detections in Bavaria, Germany, since spatial patterns of Pu agreed better with soil mapping than that of $^{137}$Cs. Similar conclusions were reached in several other studies, reporting more homogenous distributions of $^{239+240}$Pu than $^{137}$Cs in parts of Europe that were influenced by the Chernobyl accident [31, 38]. Thus, the isotopic composition variations may lead Pu isotopes to be a better choice than $^{137}$Cs for tracing soil erosion. However, the feasibility of applying $^{239,240}$Pu to soil erosion and soil C investigations still needs further testing because of its vertical migration and lateral transport by water, which may influence the quantification of soil redistribution [39, 40].

### 3.2 Natural radionuclides

$^{7}$Be is a natural radionuclide produced primarily through cosmic-ray spallation of oxygen and nitrogen nuclei in the stratosphere and troposphere. Unlike the anthropogenic radionuclides of $^{137}$Cs and $^{239,240}$Pu, wet and dry deposition of $^{7}$Be to the soil surface is continuous because of its natural origin [41]. The concentration of $^{7}$Be in the soil decreases with increases in particle size [42]. Vertical migration was low for this isotope due to its relatively short half-life (half-life = 53.3 days). Wallbrink and Murray [42] found that most of soil $^{7}$Be was accumulated in the top 20 mm regardless of soil types and surface cover conditions at two Australian sites.

Because of its short half-life, $^{7}$Be has the potential to trace short-term soil redistribution. A number of studies have used this isotope to examine short-term soil erosion after heavy rain events, providing a basis for understanding sediment transported by dispersed overland flow [41–44]. Considering the lower vertical transport of $^{7}$Be relative to $^{137}$Cs and excess $^{210}$Pb ($^{210}Pb_{ex}$), this isotope has been applied to discriminate vertical and horizontal erosion by combining with the other two isotopes [45, 46]. Ryken et al. [47] suggested correction factors related to particle size and variations in relaxation mass depth should be used to get precise estimates of $^{7}$Be-derived soil redistribution rates. Li et al. [48] successfully applied $^{7}$Be to quantify SOC sequestration changes caused by land-use and management activity.

Another natural isotope, Lead-210 ($^{210}$Pb) produced through the decay of gaseous Radon-222 ($^{222}$Rn) generated from the decay of Radium-226 ($^{226}$Ra) in the Uranium-238 ($^{238}$U) decay series. $^{226}$Ra exists in soils and rocks. Most of $^{226}$Ra decay to $^{210}$Pb in situ, which is termed supported $^{210}$Pb. Due to diffusion, a small portion of $^{226}$Ra-derived $^{222}$Rn enters to the atmosphere and subsequently introduces $^{210}$Pb to the atmosphere. This kind of $^{210}$Pb falls to land surface through wet and dry deposition and is termed unsupported or excess $^{210}$Pb ($^{210}Pb_{ex}$). The $^{210}Pb_{ex}$ is strongly absorbed by fine soil particles and transports with soil movement. $^{210}Pb_{ex}$ fallout is continuous over time, which is like $^{7}$Be, but its half-life (half-life = 22.3 years) is longer than $^{7}$Be. Therefore, $^{210}Pb_{ex}$ has the potential to penetrate deeper soil layers (to 10 cm) than $^{7}$Be [7, 23].

The use of $^{210}Pb_{ex}$ in soil and sediment studies has increased in recent decades. Due to the continuous fallout, $^{210}Pb_{ex}$ can be used to provide long-term soil redistribution rates. Meanwhile, unlike $^{137}$Cs, $^{210}Pb_{ex}$ does not have below-detection-limit problems caused by medium-lived anthropogenic fallout radionuclides. In practice, two or more different fallout radionuclides can be used to understand the soil erosion history. $^{210}Pb_{ex}$ is combined with $^{137}$Cs and/or $^{7}$Be to provide soil redistribution records in the past 100 years. Several studies suggested that $^{210}Pb_{ex}$ and $^{7}$Be can produce similar spatial patterns [45, 49], but differences resulting from different land uses could help sediment source identification [46, 50]. Increasing applications of $^{210}Pb_{ex}$ in C dynamic studies have been reported recently [10, 15, 51, 52]. Investigations have demonstrated that $^{210}Pb_{ex}$ is preferentially associated with SOC than $^{137}$Cs, due to the stronger binding to the organic matter in soils [53, 54].

### 3.3 C isotopes

The origin of soil C isotopes ($^{12}C$, $^{13}C$, and $^{14}C$) is mainly from plant litter entering the soils. During photosynthesis, plant species absorb atmospheric carbon dioxide ($CO_2$), which consists of about 98.9% of $^{12}CO_2$, 1.1% of $^{13}CO_2$, and trace amount of $^{14}CO_2$ (1 part in trillion). $^{14}C$ is a natural radionuclide created primarily by the cosmic-ray spallation of nitrogen in the troposphere and the stratosphere. $^{14}C$ is a relatively stable radioactive isotope with a half-life of 5730 years [55]. This C isotope is assimilated during plant photosynthesis and entering soil through litter fall and commonly used for age dating.

Data on the stable C isotopes $^{12}C$ and $^{13}C$ are usually reported $\delta^{13}C$ representing deviation of measured $^{13}C$ from the established natural abundance with units of parts per thousand (‰). Variations in soil $\delta^{13}C$ values are controlled primarily by carbon input from plant litter. Due to isotopic discrimination by their photosynthetic enzymes and the regulation of stomatal diffusion resistance, $\delta^{13}C$ values in plants cover a wide range. Plants with C3 photosynthesis have $\delta^{13}C$ values in a range of −22 to −32‰. Plants with C4 photosynthesis are less depleted in $^{13}C$ with higher $\delta^{13}C$ values ranging from −9 to −17‰. After plant residues entering the soil, the $\delta^{13}C$ values may change slightly due to isotope fractionation during microbial decomposition.

Soil C isotopes can effectively detect soil redistribution and reflect soil C dynamics over the landscape. Because of the isotopic variability in different plant species, soil $\delta^{13}C$ has been successfully applied to distinguish eroded soil sources and to identify soil sources from different land use types [56, 57]. Furthermore, due to the significant difference of C isotopic ($\delta^{13}C$ and $^{14}C$) values between surface soils and subsoils, scientists also utilized the isotopes to obtain a better understanding of the origins and age of eroded SOC [2, 57]. Trends of $\delta^{13}C$ and $^{14}C$ by soil depth can be used to reconstruct history of vegetation succession based on different isotopic compositions [55, 58]. Li et al. [29] used both $\delta^{13}C$ and $^{137}Cs$ to investigate soil C fate in an agricultural field under an annual crop rotation of soybean (*Glycine max* [L.] Merr.) and maize (*Zea mays* L.) which represent C3 and C4 crop types respectively. They found that only C3-derived SOC was highly correlated with soil redistribution rates. The results indicated that SOC from different plant species might have different responses to physical disturbances (i.e., erosion and deposition) and biogeochemical transformations such as SOC mineralization.

## 4. Use of isoscapes in studies of soils and sediments

The spatial and temporal features of isotopes in various environmental materials can be predicted using both landscape and biogeochemical models [59]. The term of isoscape was introduced in the 2000s to describe spatial patterns of isotopes produced by various quantitative models using several environmental variables as predictors [60, 61]. Now isoscape analyses are widely used as a tool to understand the biogeochemistry of landscapes under study. Applications of isoscapes on soils and sediments can be categorized into three groups:

1. Isotopes can be used as the baseline in modeling studies. With the help of a robust model, regional isotopes can be estimated from gridded environmental datasets. Isoscapes provide a powerful tool for probing the efficiencies of these prediction models [62];

2. Comparison of isotopic spatial patterns with observed soil organic C and nitrogen (N) is useful for investigating the impacts of soil redistribution on soil properties. Isoscapes can effectively reflect soil redistribution patterns across space, which can be used to better explain spatial variability of soil properties [29, 63, 64];

3. Soil isoscapes is also used in investigations of plant and animal isoscapes. For example, because the primary input of $^{12}C$ and $^{13}C$ to soil are from plant litter, maps of soil $\delta^{13}C$ isoscape could provide information related to the biogeography of C3 and C4 plants and their relative contributions to the atmospheric C sequestration [65].

## 5. Application of isotopic analysis in an agricultural field

### 5.1 Study area introduction

A case study is presented to demonstrate the application of isotopic analysis and isoscape modeling in investigation of soil redistribution and C dynamics in cropland. The study was conducted in an agricultural field in Walnut Creek Watershed (WCW), Iowa. The WCW is in a humid continental climatic zone with relatively flat terrain (**Figure 3a**). Representative soils are a poor-drained Nicollet (mesic Aquic Hapludolls) in the lowlands and well-drained Clarion (mesic Typic Hapludolls) in the uplands. The dominant land use type in the WCW is agriculture. More than 86% of the watershed is farmlands. Primary tillage practices in the watershed are chisel plowing in autumn and disking in spring. The directions of tillage depend on the farm management and field configuration. Most of the area followed the north-south or east-west tillage directions.

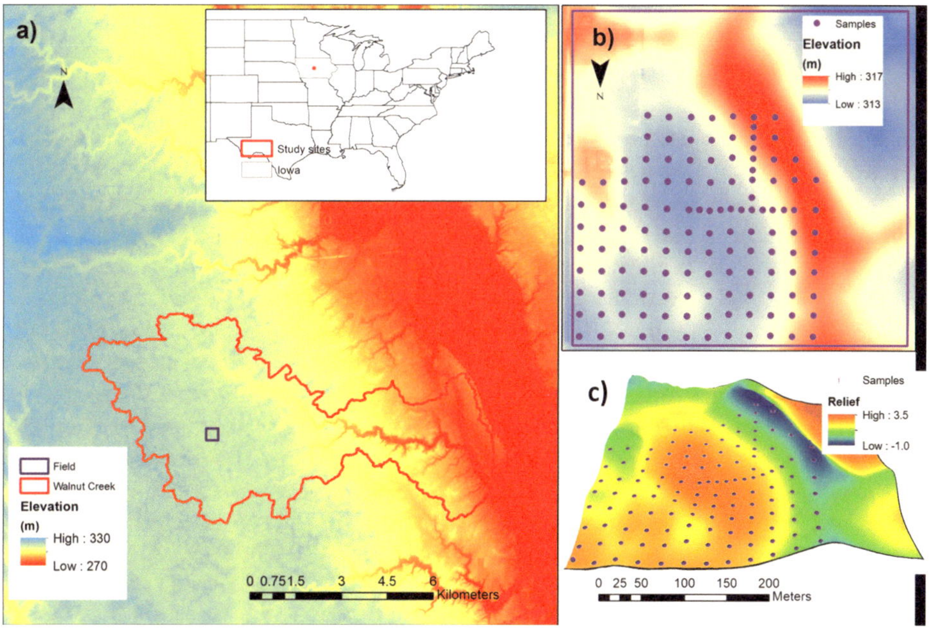


**Figure 3.**
*Elevation variability of (a) Walnut Creek Watershed (WCW) and (b) the cropland field, and (c) relief variability of the cropland field.*

The studied cropland field is about 15 ha (**Figure 3b**). The maximum elevation change within a radius of 90 m is 4.5 m (**Figure 3c**). A C4–C3 crops rotation (i.e., maize-soybean) has been adopted in this area since 1957, which make this field an appropriate testbed for a detailed C isotope study assessing impacts of soil redistribution.

## 5.2 Methods

### 5.2.1 Sampling and laboratory analyses

Isotopes of $^{137}Cs$, $^{12}C$, and $^{13}C$ were measured in soil samples. A total of 119 locations were collected at grid nodes in a 25 × 25 m grid. A 9-sample and an 11-sample transects were set at an interval of 12.5 m within the grid (**Figure 2b** and **c**). We also selected four reference sites for estimating the baseline $^{137}Cs$ inventory. Three samples were collected from the top 30 cm of soil within a 1 × 1 m quadrate using a 3.2 cm diameter push probe at each location. All soil samples were oven dried and weighed for soil bulk density calculation.

SOC content was estimated as the difference between total C that was measured by a LECO CNS 2000 elemental analyzer and calcium C estimated using soil samples ashed in a furnace at 420°C. SOC content (%) was them converted to SOC density (kg $m^{-2}$) using soil bulk density.

The $^{13}C/^{12}C$ isotope ratio was measured using an isotope-ratio mass spectrometer (Europa Scientific Ltd., Crewe, England). The ratio was then combined with the laboratory reference that was calibrated against the international PeeDee Belemnite (PDB) to estimated $\delta^{13}C$ related to PDB.

This study estimated the relative contribution of C3 and C4 crops using the following equations:

$$F_{C4} = \frac{\delta^{13}C - \delta^{13}C_{C3}}{\delta^{13}C_{C4} - \delta^{13}C_{C3}} \quad (1)$$

$$F_{C3} = 1 - F_{C4} \quad (2)$$

where $F_{C4}$ and $F_{C3}$ are contribution fraction of C4- and C3- derived SOC, respectively. $\delta^{13}CC_3$ and $\delta^{13}CC_4$ are the isotopic signatures for C3 and C4 crops. According to the literature review, the average $\delta^{13}CC_3$ and $\delta^{13}CC_4$ values for soybean and maize are −27 and −12%. Therefore, we used these two values to obtain the likelihood C3- and C4-derived SOC in this field.

To measure $^{137}Cs$, gamma-ray analysis was performed using a Canberra Genie-2000 Spectroscopy System. Original unit of $^{137}Cs$ concentration was becquerels per gram (Bq $g^{-1}$), and this unit was subsequently converted to becquerels per square meter (Bq $m^{-2}$) using soil bulk density. The baseline $^{137}Cs$ inventory estimated from reference sites and the $^{137}Cs$ inventory of sampling sites were used to calculate $^{137}Cs$-derived soil redistribution rates using the Mass Balance model I developed by Walling et al. [66]. In this study, the baseline $^{137}Cs$ inventory was 2657 Bq $m^{-2}$. Sites with higher $^{137}Cs$ inventories than the baseline were considered as depositional sites; while sites with lower $^{137}Cs$ inventories were referred as eroded sites.

Additional sampling and laboratory analyses details are given in Li et al. [29].

### 5.2.2 Historic orthophoto interpretation

Historic orthophotos in the 1950s and 2002 were obtained from the Iowa geographic map server (http://ortho.gis.iastate.edu/) to visualize soil movement. The surface soil layer was referred as mollic epipedon formed under grass vegetation that dominated this area prior to the 1850s. The mollic epipedon has a characteristic black color and contains a high amount of soil organic matter. Therefore, the black mollic epipedon can be visually identified in the historic orthophoto, serving as an indicator for soil erosion and deposition investigation after the 1850s. In this study, greyness values were derived from the orthophotos to quantify the intensity of black carbon signal related to the mollic epipedon. The values were compared to $^{137}$Cs-derived soil redistribution rates to assess the appropriateness of using $^{137}$Cs to trace soil movement in this low-relief cropland.

### 5.2.3 Terrain analyses

Topographic metrics used to describe features of the landscape were generated from the light detection and ranging (LiDAR)—derived 3 m digital elevation model (DEM). Fourteen topographic metrics were developed, including slope, profile curvature (P_Cur), plan curvature (Pl_Cur), general curvature (G_Cur), flow accumulation (FA), topographic relief [topographic relief principal components (TRPCs) and topographic relief factors (TRFAs)], positive topographic openness (PTO), downslope index (DI), flow path length (FPL), catchment area (CA), topographic wetness index (TWI), stream power index (SPI), upslope slope (UpSl), and flow length factor (LS) (**Table 1**). Before calculating topographic metrics, the DEM was filtered twice through a low pass 3-by-3 filter. Maps of slope, P_Cur, Pl_Cur, G_Cur, PTO, DI, FPL, CA, TWI, SPI, UpSl, and LS were developed by the System for Automated Geoscientific Analyses (SAGA) v. 2.2.5 [67]. FA map was generated with ArcGIS.

For TRPCs, a series of topographic relief maps with different spatial scales were generated. The topographic relief shows the elevation differences between the filtered DEM and a maximum elevation map showing a continuous surface of maximum elevation within a specific area/radius. Seven relief images were generated from seven maximum elevation maps using radiuses of 7.5, 15, 30, 45, 60, 75, and 90 m. Principal component analysis (PCA) was conducted to convert the seven reliefs to seven independent components, and the first two topographic relief principal components (TRPC1 and TRPC2) were selected for further analysis. Similarly, varimax rotated Factor Analysis (FAn) was used and the first two TRFA1 and TRFA2 were selected. The detailed topographic processing can be found in Li et al. [68].

### 5.2.4 Statistical analysis and model calibration

Duncan's multiple range tests ($p \leq 0.05$) were applied to test the mean differences of soil texture, soil SOC density, $\delta^{13}$C, $^{137}$Cs inventory, $^{137}$Cs-derived soil redistribution rates, and likelihood C3- and C4-derived SOC density between eroded and depositional sites. Topographic metrics of the 128 sampling locations were extracted from the DEM-derived topographic metric maps. Because some of the topographic metrics were highly correlated, PCA and varimax rotated FAn were applied to convert to mutually independent topographic combinations to reduce errors caused by collinearity between the metrics. The first six principal components (TPCs) and the first six factors (TFAs) were selected to develop isoscape models.

Models including the multiple linear regression (MLR) combined with principal component analysis (MLR-PCA) and MLR combined with factor analysis (MLR-FAn) models for SOC, $^{137}$Cs, and $\delta^{13}$C were developed using stepwise linear regression with the "leave-one-out" cross-validation. We randomly selected 70% samples

| Variables | Definition | Generating method |
|---|---|---|
| Slope (radian) | An angle between a tangent and a horizontal plane at a given point | Slope, aspect, curvature module, SAGA |
| P_Cur ($m^{-1}$) | Curvature of the surface in the direction of the steepest slope | |
| Pl_Cur ($m^{-1}$) | Curvature in a horizontal plane | |
| G_Cur ($m^{-1}$) | Curvature of the surface itself | |
| FA ($m^2$) | Land area that contributes surface water to an area in which water accumulates | ArcGIS |
| TRPC and TRFA | Topographic relief is elevation difference between the highest point over an area and a given location. TRPC/TRFA is topographic relief principal component/Topographic relief factor and is generated by several reliefs that are at different spatial-scales using principal component analysis/factor analysis | ArcGIS |
| PTO (radian) | An angular measure of the relation between surface relief and horizontal distance | Topographic openness module, SAGA |
| DI (radian) | Head differences along a flow path | Downslope distance gradient module, SAGA |
| FPL (m) | Maximum distance of water flow to a point in the catchment | Flow path length module, SAGA |
| CA ($m^2$) | Area draining to catchment outlet | SAGA wetness index module, SAGA |
| TWI | Frequencies and duration of saturated conditions | |
| SPI | Erosive power of overland flow | Stream power index module, SAGA |
| UpSl (radian) | Mean slope of upslope area | LS-factor (field based) module, SAGA |
| LS | Erosive power of the terrain | |

*P_Cur, Pl_Cur, and G_Cur are profile curvature, plan curvature and general curvature, respectively; TRPC and TRFA are topographic relief principal components and topographic relief factors, respectively; PTO is positive topographic openness; DI is downslope index; CA is catchment area; TWI is topographic wetness index; SPI is stream power index; Upsl is upslope slope; LS is slope length factor.*

**Table 1.**
*Definitions and generating methods of selected topographic metrics.*

as training dataset used for model calibration and 30% samples as testing dataset for model validation. Three criteria were used to evaluate model efficiencies. The three criteria are the adjusted coefficient of determination ($R^2_{adj}$), the Nash-Sutcliffe efficiency (NSE), and ratio of the root mean square error (RMSE) to the standard deviation of measured data (RSR). Usually, the model performance is considered satisfactory when NSE is larger than 0.5 and RSR is smaller than 0.7 [69].

## 5.3 Results and discussion

### *5.3.1 Soil redistribution impacts on C dynamics*

Distribution of mollic epipedon, characterized by presence of black soil organic matter, reflected historical soil movement (**Figure 4**). The top-layer black soil presumably blanketed the prairie landscape and was then transported by water and tillage from eroded upslope to downslope locations. This phenomenon can be observed from the higher mollic epipedon in the concave than the convex locations in the 1950s and likely more consolidated by 2002. The significant correlations between greyness values and $^{137}$Cs-derived soil redistribution rate suggested that $^{137}$Cs inventory can effectively reflect the soil redistribution process in this field.

The cropland experienced a general soil export as indicated by a negative mean $^{137}$Cs-derived soil redistribution rate ($-0.51 \pm 2.09$ kg m$^{-2}$ year$^{-1}$, **Table 2**). Specifically, about 81 locations were identified as eroded sites with negative soil redistribution rates (mean: $-1.72 \pm 1.38$ kg m$^{-2}$ year$^{-1}$) and 47 locations were depositional sites with positive soil redistribution rates (mean: $1.57 \pm 1.32$ kg m$^{-2}$ year$^{-1}$). SOC density exhibited a similar spatial pattern as soil redistribution rates with a higher mean value at depositional than eroded sites. The mean SOC density at depositional sites was about 1.70 times of that at eroded sites.

A strong and significant correlation was observed between $^{137}$Cs-derived redistribution rates and SOC density ($r^2 = 0.667$, **Figure 5a**), which demonstrated that $^{137}$Cs inventories can capture spatial patterns of SOC. The positive relationship between soil redistribution and SOC density is mainly caused by the preferential removal of C during soil erosion. The SOC is primarily concentrated on the topsoil layer and is prone to transport through runoff, wind, and tillage activities due to its low density. Silt and clay density that were enriched in SOC showed lower values at eroded sites than depositional sites could also provide evidence for the preferential movement of low-density particles [29].

Note that SOC mineralization and dynamic replacement can complicate impacts of soil redistribution on SOC. On the one hand, SOC mineralization highly depends on soil moisture conditions. Impacted by soil texture, soil moisture is commonly lower at upslope areas than low-lying areas. Our study area also showed similar spatial patterns in soil texture over the landscape [29]. The decreased fine particles

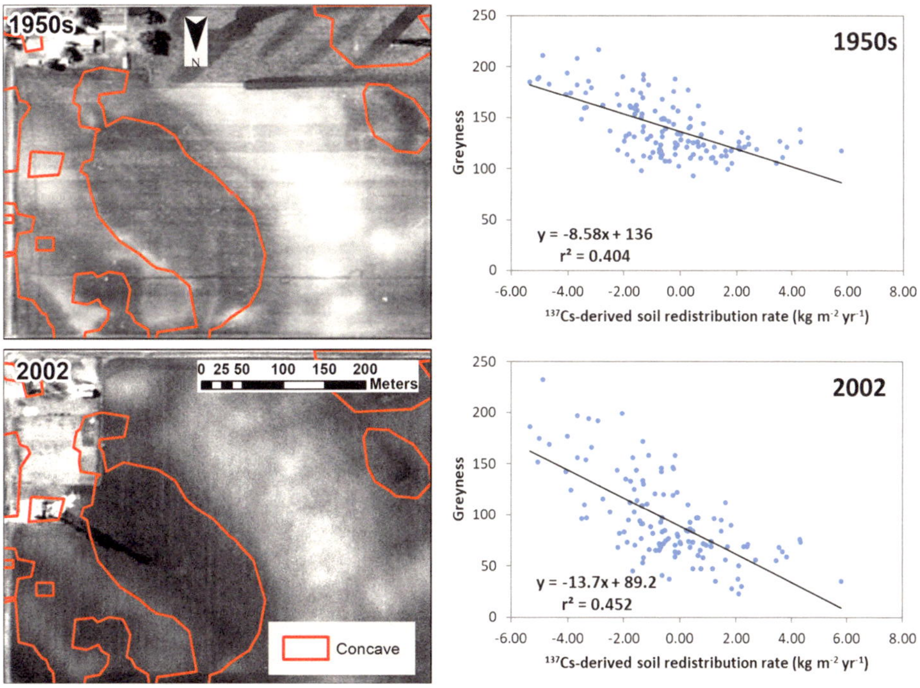


**Figure 4.**
*Historic orthophotos and bivariate relationships between greyness and $^{137}$Cs inventory in the 1950s and 2002.*

| | N | SOC | $\delta^{13}C$ | $^{137}Cs$ | SR | C3 | C4 |
|---|---|---|---|---|---|---|---|
| Erosion | 81 | 759 (246) b† | −19.6 (2.05)a | 1791 (532)b | −1.72 (1.38)b | 396 (202)b | 363 (131)a |
| Deposition | 47 | 1292 (259)a | −21.9 (1.52)b | 3547 (769)a | 1.57 (1.32)a | 852.2 (226)a | 440 (152)a |
| All | 128 | 956 (359) | −20.4 (2.16) | 2435 (1056) | −0.51 (2.09) | 565 (305) | 391 (143) |

*N is the number of samples.*
*†Letters (a and b) estimate based on Duncan's multiple range tests. There are no significant ($p < 0.05$) differences for a parameter with the same letter.*

**Table 2.**
*Means (standard deviations) of soil organic carbon (SOC, kg $m^{-2}$), isotopic signature ($\delta^{13}C$, %), Cesium-137 ($^{137}Cs$) inventory (Bq $m^{-2}$), $^{137}Cs$-derived soil redistribution (SR, kg $m^{-2}$ $year^{-1}$), and likelihood C3- and C4-derived SOC density (kg $m^{-2}$) in the cropland field.*

(clay and silt) can reduce soil water retention, decrease aggregate stability, and increase oxygen concentration, accelerating SOC mineralization and further exacerbate SOC depletion in eroded areas [12]. On the other hand, the depleted SOC pool can be dynamically replaced by the continually deposited SOC from litter decomposition of above- and below-ground biomass. The net sequestration of carbon is not always linear over time because rates of replaced carbon at eroded sites depend on management history that may vary with time [18].

Values of $\delta^{13}C$ showed opposite spatial patterns with SOC and $^{137}Cs$-derived soil redistribution rates with a higher mean value of $\delta^{13}C$ at eroded than depositional sites. The mean $\delta^{13}C$ value at the eroded sites was −19.6 ± 2.05%; while the $\delta^{13}C$ value at the depositional sites was −21.9 ± 1.52%. The higher mean $\delta^{13}C$ value at eroded sites indicated that the eroded sites were less depleted for $^{13}C$ and exhibited stronger C4 vegetation characteristics than depositional sites. The $\delta^{13}C$ value variability over this space possibly resulted from dynamic replacement. Our study area was dominated by prairie vegetation with strong C3 vegetation characteristics before the 1850s. After that, C4-derived SOC was introduced as a new SOC formation into the soil due to the widespread maize cultivation [70]. The new soil C (C4-derived SOC) replaced 50% of native soil C (C3-derived SOC) at eroded sites and 30% at depositional sites till 2002 when sampled.

According to the calculations of the relative contributions of C3 and C4 crops to the SOC, the C3-derived SOC density showed a significant difference between eroded (396 ± 202 kg $m^{-2}$) and depositional sites (852 ± 226 kg $m^{-2}$) while the C4-derived SOC density varied insignificantly (**Table 2**, eroded: 363 ± 131 kg $m^{-2}$; depositional: 440 ± 152 kg $m^{-2}$). C3-derived SOC density was positively related to $^{137}Cs$-derived soil redistribution rates with a coefficient of determination of 0.65 (**Figure 5b**). In contrast, C4-derived SOC density was not strongly related

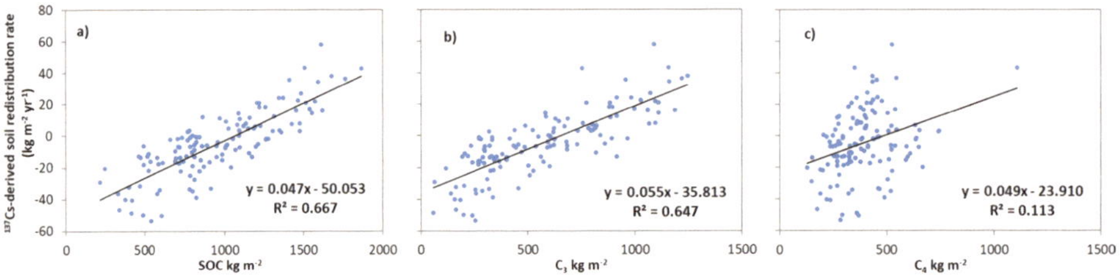


**Figure 5.**
*Bivariate relationships between $^{137}Cs$-derived soil redistribution rates and (a) SOC density, (b) C3- and (c) C4-derived SOC density.*

to soil redistribution rates (**Figure 5c**). The different responses to soil erosion may be related to chemical compounds contained in the C3- and C4-derived SOC. Generally, liable compounds that are high in C4 crop residues are less depleted in $^{13}C$ than lignin-based compounds that are high in C3 crop residues [71]. Therefore, C4-derived SOC may largely be found in the mollic epipedon and be significantly affected by C mineralization, resulting in a decreased correlation with patterns of soil movement. In contrast, the more recalcitrant SOC in soil with strong C3 vegetation characteristics would be less influenced by C mineralization because it contains higher amount of stable lignin-based compounds. Therefore, the variability in C3 signature of soil C would more strongly correlate with patterns of soil redistribution.

### *5.3.2 Isoscape model development*

Based on the derived topographic metrics, the first six TPCs and six TFAs were selected to construct independent topographic variables. The metrics with the highest absolute loadings for the first six TPCs were TWI, TRPC2, FA, FPL, DI, and CA. Therefore, the TPCs 1, 2, 3, 4, 5, and 6 were associated with soil wetness, flow acceleration, runoff volume, flow velocity, downslope dispersal, and water volume, respectively [72]. The highest absolute loadings in the six TFAs were Upsl, G_Cur, Pl_Cur, FPL, CA, and DI, which were associated with upslope flow velocity, flow acceleration, flow convergence and flow divergence, flow velocity, water volume, and downslope dispersal, respectively.

SOC and isoscape models of $\delta^{13}C$ and $^{137}Cs$ were developed using MLR-PCA and MLR-FAn (**Table 3**). Both types of models (MLR-PCA and MLR-FAn) presented satisfactory efficiencies during model calibration, but more independent variables were selected in MLR-FAn models. For SOC simulation, the MLR-PCA model selected four components with a $R^2_{adj}$ of 0.668, an NSE of 0.682, and an RSR of 0.563. The MLR-FAn model exhibited higher $R^2_{adj}$ and NSE values and a lower RSE value than MLR-PCA SOC model. In contrast, for $^{137}Cs$ inventory and $\delta^{13}C$ simulations, similar $R^2_{adj}$ values were observed between the two types of models, but the MLR-FAn models showed higher NSE and RSR values, which suggested that the $^{137}Cs$ and $\delta^{13}C$ values derived from MLR-FA models were less deviated from the corresponding observed values.

| | Model | $R^2_{adj}$ | NSE | RSR |
|---|---|---|---|---|
| MLR-PCA | | | | |
| SOC | 949-119TPC1 + 86.9TPC5 + 21.4TPC2 + 37.2TPC4† | 0.668 | 0.682 | 0.563 |
| $^{137}Cs$ | 2466-327TPC1 + 240TPC5 | 0.597 | 0.606 | 0.628 |
| $\delta^{13}C$ | −20.3 + 0.677TPC1 + 0.250TPC3 + 0.181TPC2 | 0.614 | 0.626 | 0.611 |
| MLR-FAn | | | | |
| SOC | 950 + 189TFA4-147TFA2-112TFA1 + 88.9TFA6 + 82.1TFA3 + 50.2TFA5 | 0.663 | 0.686 | 0.560 |
| $^{137}Cs$ | 2459 + 447TFA4 + 306TFA3-346TFA1 + 316TFA6-301TFA2 + 150TFA5 | 0.587 | 0.615 | 0.621 |
| $\delta^{13}C$ | −20.3 + 1.26TFA1 + 0.717TFA2-0.741TFA4-0.406TFA6-0.317TFA5 | 0.614 | 0.635 | 0.604 |

†*The order of variables is based on the stepwise selection steps. TPC and TFA are topographic principal component and topographic factor, respectively.*

**Table 3.**
*Topography-based models for soil organic carbon (SOC), cesium-137 ($^{137}Cs$), and isotopic signature ($\delta^{13}C$) in the cropland field.*

Although both types of models reasonably matched up with observations during model calibration, the MLR-PCA models had better performance in predictions than the MLR-FAn models when applied to the test dataset. Predicted SOC explained 62.0% variability in observed SOC using MLR-PCA model. The values of NSE and RSR were 0.612 and 0.622, respectively. MLR-PCA $^{137}Cs$ and $\delta^{13}C$ models also had satisfactory performance with NSE values larger than 0.5 and RSR values smaller than 0.7. The $R^2$ for $^{137}Cs$ and $\delta^{13}C$ prediction were 0.713 and 0.509, respectively. In contrast, all the MLR-FAn models presented lower efficiencies in predicting target variables. Especially for the $\delta^{13}C$, the NSE value was smaller than 0.5 and RSR value was larger than 0.7 when compared MLR-FAn predictions with observations.

The lower efficiencies of the MLR-FAn models than the MLR-PCA models may be caused by model over-fitting. PCA considers all the variance in the independent variables, including unique, error and shared variance during synthetic variable construction; while FAn only considered and presented the shared variance in the factor matrix. Therefore, differences exist during development of MLR-PCA and MLR-FAn models. In this case study, although these two types of models had similar performance during model calibration using training datasets, increased number of parameters in MLR-FAn models may increase model error using validation datasets with general decrease in model stability, leading to low accuracies during extrapolating prediction points to external sample sets [73].

The spatial patterns of SOC density, $^{137}Cs$ inventory, and $\delta^{13}C$ were generated using MLR-PCA models (**Figure 6**). SOC density showed a similar spatial variability as $^{137}Cs$ inventories, which is consistent with the strong correlations between the SOC density and $^{137}Cs$-derived soil redistribution rate (**Figure 5a**). Both variables had high values in depressions and low values in ridge and sloping areas. For $\delta^{13}C$, an opposite pattern was observed with low values in concave landscapes and high values in convex landscapes.

The high model efficiencies demonstrated the feasibility of using topography-based isoscape MLR-PCA models for investigating the spatial variability of isotopes. Although limitations may exist due to unaccounted topographic variables and environmental impactors, such as variance in tillage activities, the DEM-derived topography-based models provide a cost-effective method for isoscape mapping, effectively reflecting

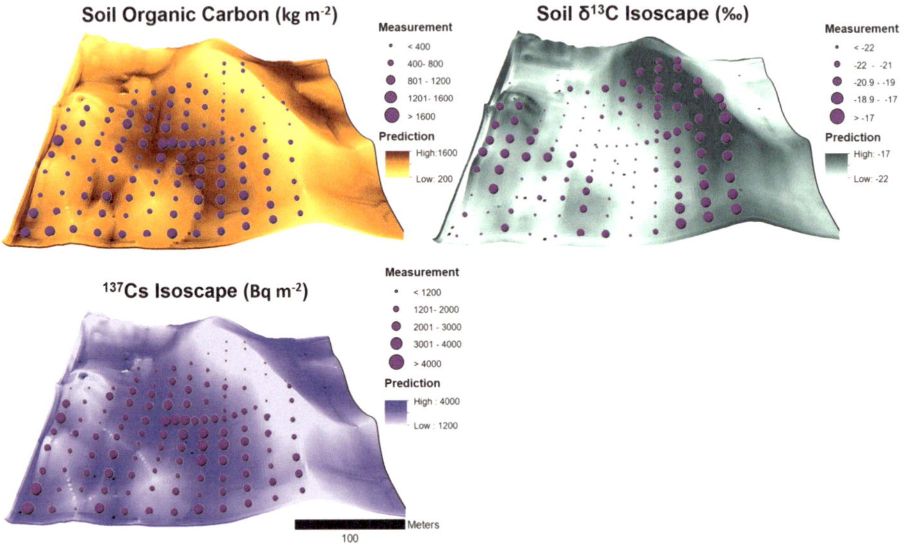


**Figure 6.**
*Spatial patterns of SOC density, $^{137}Cs$ inventory, and $\delta^{13}C$ in the study field.*

redistribution patterns of soil and spatial patterns of SOC over the landscape [28, 74, 75]. These models can also benefit investigations in regions with limited access to ground measurements. The recently increased availability of the fine-resolution LiDAR data can further improve the model applicability, allowing for scaling of in situ isotopic simulations from field scale to isoscape mapping at watershed or regional scales.

## 6. Conclusions

Soil isotopes can effectively trace soil redistribution and SOC dynamics. This chapter reviewed the application of natural ($^{7}Be$, $^{210}Pb$), anthropogenic fallout radionuclides ($^{137}Cs$, $^{239,240}Pu$), and C isotopes ($^{12,13,14}C$) in understanding soil erosion and deposition at different spatial and temporal scales. The case study demonstrated that $^{137}Cs$ and C isotopes could be employed to understand soil movement and C dynamics. SOC density showed high consistency with $^{137}Cs$-derived soil redistribution rate, suggesting significant spatial impacts of soil movement on SOC. $\delta^{13}C$ provided further support of the importance of dynamic replacement on soil C dynamics in this area. Topography-based isoscape models were developed and effectively reconstructed the spatial variability in $^{137}Cs$ inventory and $\delta^{13}C$ over the landscape. The isoscape maps of $^{137}Cs$ and $\delta^{13}C$ were in general agreement with the spatial pattern of SOC density. Based on such results, we conclude that isotopic and isoscape analysis could provide valuable insights into soil movement and C studies.

## Acknowledgements

This research was supported by the USDA Natural Resources Conservation Service in association with the Wetland Component of the National Conservation Effects Assessment Project (NRCS 67-3A75-13-177).

## Conflict of interest

No conflict of interest exists relative to information presented in this chapter.

## Author details

Xia Li, Gregory McCarty* and Sangchul Lee
Hydrology and Remote Sensing Laboratory, USDA-ARS, Beltsville, MD, USA

*Address all correspondence to: greg.mccarty@usda.gov

## References

[1] Lal R. Soil erosion and the global carbon budget. Environment International. 2003;**29**(4):437-450. DOI: 10.1016/S0160-4120(02)00192-7

[2] Nadeu E, Berhe AA, De Vente J, Boix-Fayos C. Erosion, deposition and replacement of soil organic carbon in Mediterranean catchments: A geomorphological, isotopic and land use change approach. Biogeosciences. 2012;**9**(3):1099-1111. DOI: 10.5194/bg-9-1099-2012

[3] Larson WE, Pierce FJ, Dowdy RH. The threat of soil erosion to long-term crop production. Science. 1983;**219**(4584):458-465. DOI: 10.1126/science.219.4584.458

[4] Amore E, Modica C, Nearing MA, Santoro VC. Scale effect in USLE and WEPP application for soil erosion computation from three Sicilian basins. Journal of Hydrology. 2004;**293**(1-4):100-114. DOI: 10.1016/j.jhydrol.2004.01.018

[5] Schumacher JA, Kaspar TC, Ritchie JC, Schumacher TE, Karlen DL. Identifying spatial patterns of erosion for use in precision conservation. Journal of Soil and Water Conservation. 2005;**60**(6):355-362

[6] Du P, Walling DE. Using $^{137}Cs$ measurements to investigate the influence of erosion and soil redistribution on soil properties. Applied Radiation and Isotopes. 2011;**69**(5):717-726. DOI: 10.1016/j.apradiso.2011.01.022

[7] Matisoff G, Whiting PJ. Measuring soil erosion rates using natural ($^{7}Be$, $^{210}Pb$) and anthropogenic ($^{137}Cs$, $^{239,240}Pu$) radionuclides. In: Handbook of Environmental Isotope Geochemistry. Berlin: Spring; 2011. pp. 487-519. DOI: 10.1007/978-3-642-10637-8

[8] Le Quéré C, Andrew RM, Friedlingstein P, Sitch S, Hauck J, Pongratz J, et al. Global carbon budget 2018. Earth System Science Data. 2018;**10**:2141-2194. DOI: 10.5194/essd-8-605-2016

[9] Kosmas C, Gerontidis S, Marathianou M, Detsis B, Zafiriou T, Nan Muysen W, et al. The effects of tillage displaced soil on soil properties and wheat biomass. Soil and Tillage Research. 2001;**58**(1-2):31-44. DOI: 10.1016/S0167-1987(00)00175-6

[10] Li Y, Zhang QW, Reicosky DC, Lindstrom MJ, Bai LY, Li L. Changes in soil organic carbon induced by tillage and water erosion on a steep cultivated hillslope in the Chinese loess plateau from 1898-1954 and 1954-1998. Journal of Geophysical Research-Biogeosciences. 2007;**112**(1):1-10. DOI: 10.1029/2005JG000107

[11] Martinez-Mena M, Lopez J, Almagro M, Boix-Fayos C, Albaladejo J. Effect of water erosion and cultivation on the soil carbon stock in a semiarid area of South-East Spain. Soil and Tillage Research. 2008;**99**(1):119-129. DOI: 10.1016/j.still.2008.01.009

[12] Wang X, Cammeraat ELH, Cerli C, Kalbitz K. Soil aggregation and the stabilization of organic carbon as affected by erosion and deposition. Soil Biology and Biochemistry. 2014;**72**:55-65. DOI: 10.1016/j.soilbio.2014.01.018

[13] Polyakov VO, Lal R. Soil erosion and carbon dynamics under simulated rainfall. Soil Science. 2004;**169**(8):590-599. DOI: 10.1097/01.ss.0000138414.84427.40

[14] Van Oost K, Govers G, De Gryze S, Six J, Harden JW, Ritchie JC, et al. The impact of agricultural soil erosion

on the global carbon cycle. Science. 2007;**318**(5850):626-629

[15] McCarty GW, Pachepsky Y, Ritchie J. Impact of sedimentation on wetland carbon sequestration in an agricultural watershed. Journal of Environmental Quality. 2009;**38**(2):804-813. DOI: 10.2134/jeq2008.0012

[16] McCarty GW, Ritchie JC. Impact of soil movement on carbon sequestration in agricultural ecosystems. Environmental Pollution. 2002;**116**(3):423-430. DOI: 10.1016/S0269-7491(01)00219-6

[17] Berhe AA, Harden JW, Torn MS, Harte J. Linking soil organic matter dynamics and erosion-induced terrestrial carbon sequestration at different landform positions. Journal of Geophysical Research – Biogeosciences. 2008;**113**(4):1-12. DOI: 10.1029/2008JG000751

[18] Harden JW, Sharpe JM, Parton WJ, Ojima DS, Fries TL, Huntington TG, et al. Dynamic replacement and loss of soil carbon on eroding cropland. Global Biogeochemical Cycles. 1999;**13**(4):885-901. DOI: 10.1029/1999GB900061

[19] Stallard RF. Terrestrial sedimentation and the carbon cycle: Coupling weathering and erosion to carbon burial. Global Biogeochemical Cycles. 1998;**12**(2):231-257. DOI: 10.1029/98GB00741

[20] Lal R. Soil erosion and carbon dynamics. Soil and Tillage Research. 2005;**81**(2):137-142. DOI: 10.1016/j.still.2004.09.002

[21] Van Oost K, Govers G, Quine TA, Heckrath G, Olesen JE, De Gryze S, et al. Landscape-scale modeling of carbon cycling under the impact of soil redistribution: The role of tillage erosion. Global Biogeochemical Cycles. 2005;**19**(4):1-13. DOI: 10.1029/2005GB002471

[22] Steinhauser G, Brandl A, Johnson TE. Comparison of the Chernobyl and Fukushima nuclear accidents: A review of the environmental impacts. Science of the Total Environment. 2014;**470-471**:800-817. DOI: 10.1016/j.scitotenv.2013.10.029

[23] Doering C, Akber R, Heijnis H. Vertical distributions of $^{210}$Pb excess, $^{7}$Be and $^{137}$Cs in selected grass covered soils in Southeast Queensland, Australia. Journal of Environmental Radioactivity. 2006;**87**(2):135-147. DOI: 10.1016/j.jenvrad.2005.11.005

[24] Schuller P, Ellies A, Kirchner G. Vertical migration of fallout $^{137}$Cs in agricultural soils from southern Chile. Science of the Total Environment. 1997;**193**(3):197-205. DOI: 10.1016/S0048-9697(96)05338-7

[25] Ramzaev V, Barkovsky A. Vertical distribution of $^{137}$Cs in grassland soils disturbed by moles (*Talpa europaea* L.). Journal of Environmental Radioactivity. 2018;**184-185**:101-108. DOI: 10.1016/j.jenvrad.2018.01.011

[26] Mabit L, Bernard C. Relationship between soil $^{137}$Cs inventories and chemical properties in a small intensively cropped watershed. Comptes Rendus de l'Académie des Sciences-Series IIA-Earth and Planetary Science. 1998;**327**(8):527-532. DOI: 10.1016/S1251-8050(99)80034-2

[27] Ritchie JC, McCarty GW, Venteris ER, Kaspar TC. Soil and soil organic carbon redistribution on the landscape. Geomorphology. 2007;**89**:163-171. DOI: 10.1016/j.geomorph.2006.07.021

[28] Li X, McCarty GW, Karlen DL, Cambardella CA. Topographic metric

predictions of soil redistribution and organic carbon in Iowa cropland fields. Catena. 2018;**160**:222-232. DOI: 10.1016/j.catena.2017.09.026

[29] Li X, McCarty GW, Karlen DL, Cambardella CA, Effland W. Soil organic carbon and isotope composition response to topography and erosion in Iowa. Journal of Geophysical Research – Biogeosciences. 2018;**123**:3649-3667. DOI: 10.1029/2018JG004824

[30] Young CJ, Liu S, Schumacher JA, Schumacher TE, Kaspar TC, McCarty GW, et al. Evaluation of a model framework to estimate soil and soil organic carbon redistribution by water and tillage using $^{137}$Cs in two U.S. Midwest agricultural fields. Geoderma. 2014;**232**:437-448. DOI: 10.1016/j.geoderma.2014.05.019

[31] Alewell C, Meusburger K, Juretzko G, Mabit L, Ketterer ME. Suitability of $^{239+240}$Pu and $^{137}$Cs as tracers for soil erosion assessment in mountain grasslands. Chemosphere. 2014;**103**:274-280. DOI: 10.1016/j.chemosphere.2013.12.016

[32] Mabit L, Benmansour M, Abril JM, Walling DE, Meusburger K, Iurian AR, et al. Fallout $^{210}$Pb as a soil and sediment tracer in catchment sediment budget investigations: A review. Earth-Science Reviews. 2014;**138**:335-351. DOI: 10.1016/j.earscirev.2014.06.007

[33] Agapkina GI, Tikhomirov FA, Shcheglov AI, Kracke W, Bunzl K. Association of Chernobyl-derived Pu-239+240, Am-241, Sr-90 and Cs-137 with organic matter in the soil solution. Journal of Environmental Radioactivity. 1995;**29**(3):257-269. DOI: 10.1016/0265-931x(95)00023-4

[34] Van Pelt RS, Ketterer ME. Use of anthropogenic radioisotopes to estimate rates of soil redistribution by wind II: The potential for future use of $^{239+240}$Pu. Aeolian Research. 2013;**9**:103-110. DOI: 10.1016/j.aeolia.2013.01.004

[35] Arata L, Alewell C, Frenkel E, A'Campo-Neuen A, Iurian AR, Ketterer ME, et al. Modelling deposition and erosion rates with RadioNuclides (MODERN)—Part 2: A comparison of different models to convert $^{239+240}$Pu inventories into soil redistribution rates at unploughed sites. Journal of Environmental Radioactivity. 2016;**162-163**:97-106. DOI: 10.1016/j.jenvrad.2016.05.009

[36] Calitri F, Sommer M, Norton K, Temme A, Brandová D, Portes R, et al. Tracing the temporal evolution of soil redistribution rates in an agricultural landscape using $^{239+240}$Pu and $^{10}$Be. Earth Surface Processes and Landforms. 2019;**44**:1783-1798. DOI: 10.1002/esp.4612

[37] Schimmack W, Auerswald K, Bunzl K. Estimation of soil erosion and deposition rates at an agricultural site in Bavaria, Germany, as derived from fallout radiocesium and plutonium as tracers. Die Naturwissenschaften. 2002;**89**(1):43-46. DOI: 10.1007/s00114-001-0281-z

[38] Alewell C, Pitois A, Meusburger K, Ketterer M, Mabit L. $^{239+240}$Pu from "contaminant" to soil erosion tracer: Where do we stand? Earth-Science Reviews. 2017;**172**:107-123. DOI: 10.1016/j.earscirev.2017.07.009

[39] Schimmack W, Auerswald K, Bunzl K. Can $^{239+240}$Pu replace $^{137}$Cs as an erosion tracer in agricultural landscapes contaminated with Chernobyl fallout? Journal of Environmental Radioactivity. 2001;**53**(1):41-57. DOI: 10.1016/S0265-931X(00)00117-X

[40] Meusburger K, Mabit L, Ketterer M, Park JH, Sandor T, Porto P, et al. A multi-radionuclide approach to evaluate the suitability of $^{239+240}$Pu as soil erosion tracer. Science of the Total

Environment. 2016;**566-567**:1489-1499. DOI: 10.1016/j.scitotenv.2016.06.035

[41] Walling DE, He Q, Blake W. Use of $^{7}$Be and $^{137}$Cs measurements to document short- and medium-term rates of water-induced soil erosion on agricultural land. Water Resources Research. 1999;**35**(12):3865-3874

[42] Wallbrink PJ, Murray AS. Distribution and variability of $^{7}$Be in soils under different surface cover conditions and its potential for describing soil redistribution processes. Water Resources Research. 1996;**32**(2):467-476. DOI: 10.1029/95WR02973

[43] Sepulveda A, Schuller P, Walling DE, Castillo A. Use of $^{7}$Be to document soil erosion associated with a short period of extreme rainfall. Journal of Environmental Radioactivity. 2008;**99**(1):35-49. DOI: 10.1016/j.jenvrad.2007.06.010

[44] Walling DE, Schuller P, Zhang Y, Iroumé A. Extending the timescale for using Beryllium 7 measurements to document soil redistribution by erosion. Water Resources Research. 2009;**45**(2):1-13. DOI: 10.1029/2008WR007143

[45] Hancock GJ, Wilkinson SN, Hawdon AA, Keen RJ. Use of fallout tracers $^{7}$Be, $^{210}$Pb and $^{137}$Cs to distinguish the form of sub-surface soil erosion delivering sediment to rivers in large catchments. Hydrological Processes. 2014;**28**(12):3855-3874. DOI: 10.1002/hyp.9926

[46] Matisoff G, Bonniwell EC, Whiting PJ. Soil erosion and sediment sources in an Ohio watershed using Beryllium-7, Cesium-137, and Lead-210. Journal of Environmental Quality. 2014;**31**(1):54-61. DOI: 10.2134/jeq2002.5400

[47] Ryken N, Vanden Nest T, Al-Barri B, Blake W, Taylor A, Bodé S, et al. Soil erosion rates under different tillage practices in Central Belgium: New perspectives from a combined approach of rainfall simulations and $^{7}$Be measurements. Soil and Tillage Research. 2018;**179**:29-37. DOI: 10.1016/j.still.2018.01.010

[48] Li Y, Yu H, Chappell A, Zhou N, Funk R. How much soil organic carbon sequestration is due to conservation agriculture reducing soil erosion? Soil Research. 2014;**52**(7):717-726. DOI: 10.1071/sr14078

[49] Whiting PJ, Matisoff G, Fornes W, Soster FM. Suspended sediment sources and transport distances in the Yellowstone River basin. Geological Society of America Bulletin. 2005;**117**(3-4):515-529. DOI: 10.1130/B25623.1

[50] Wallbrink PJ, Murray AS, Olley JM. Determining sources and transit times of suspended sediment in the Murrumbidgee River, New South Wales, Australia, using fallout $^{137}$Cs and $^{210}$Pb. Water Resources Research. 1998;**34**(4):879-887

[51] Wieder RK. Past, present, and future peatland carbon balance: An empirical model based on $^{210}$Pb-dated cores. Ecological Applications. 2006;**11**(2):327-342. DOI: 10.2307/3060892

[52] Braakhekke MC, Wutzler T, Beer C, Kattge J, Schrumpf M, Ahrens B, et al. Modeling the vertical soil organic matter profile using Bayesian parameter estimation. Biogeosciences. 2013;**10**(1):399-420. DOI: 10.5194/bg-10-399-2013

[53] Teramage MT, Onda Y, Kato H, Wakiyama Y, Mizugaki S, Hiramatsu S. The relationship of soil organic carbon to $^{210}$Pb and $^{137}$Cs during surface soil erosion in a hillslope forested environment. Geoderma. 2013;**192**(1):59-67. DOI: 10.1016/j.geoderma.2012.08.030

[54] Dörr H, Münnich KO. Downward movement of soil organic matter and its influence on trace-element transport ($^{210}Pb$, $^{137}Cs$) in the soil. Radiocarbon. 1989;**31**(3):655-663. DOI: 10.1017/s003382220001225x

[55] Follett RF, Kimble J, Leavitt SW, Pruessner E. Potential use of soil C isotope analyses to evaluate paleoclimate. Soil Science. 2004;**169**(7):471-488. DOI: 10.1097/01.ss.0000135169.45251.63

[56] Fox JF, Papanicolaou AN. The use of carbon and nitrogen isotopes to study watershed erosion processes. Journal of the American Water Resources Association. 2007;**43**(4):1047-1064. DOI: 10.1111/j.1752-1688.2007.00087.x

[57] Hancock G, Revill A. Land-Use and Erosion Source Discrimination of Soil and Carbon Sources to the Logan and Albert Rivers Using Compound Specific Isotope Analysis [Internet]. 2011. Available from: https://pdfs.semanticscholar.org/5d40/96f85df22699 4a0f4e990f46e088ed89f188.pdf

[58] Paul EA, Follett RF, Leavitt SW, Halvorson A, Peterson GA, Lyon DJ. Radiocarbon dating for determination of soil organic matter pool sizes and dynamics. Soil Science Society of America Journal. 1997;**61**:1058-1067

[59] Bowen GJ. Isoscapes: Spatial pattern in isotopic biogeochemistry. Annual Review of Earth and Planetary Sciences. 2010;**38**:161-187. DOI: 10.1146/annurev-earth-040809-152429

[60] West JB, Sobek A, Ehleringer JR. A simplified GIS approach to modeling global leaf water isoscapes. PLoS One. 2008;**3**(6):e2447. DOI: 10.1371/journal.pone.0002447. Chave J, editor

[61] West JB, Bowen GJ, Dawson TE, Tu KP. Understanding Movement, Pattern, and Process, on Earth through Isotope Mapping. Dordrecht/Heidelberg/London/New York: Springer; 2009. 487p

[62] Powell RL, Yoo E-H, Still CJ. Vegetation and soil Carbon-13 isoscapes for South America: Integrating remote sensing and ecosystem isotope measurements. Ecosphere. 2012;**3**(11):109. DOI: 10.1890/es12-00162.1

[63] Li QY, Fang HY, Sun LY, Cai QG. Using the $^{137}Cs$ technique to study the effect of soil redistribution on soil organic carbon and total nitrogen stocks in an agricultural catchment of Northeast China. Land Degradation & Development. 2014;**25**(4):350-359. DOI: 10.1002/ldr.2144

[64] Quijano L, Gaspar L, Navas A. Spatial patterns of SOC, SON, $^{137}Cs$ and soil properties as affected by redistribution processes in a Mediterranean cultivated field (Central Ebro Basin). Soil and Tillage Research. 2016;**155**:318-328. DOI: 10.1016/j.still.2015.09.007

[65] Ladd B, Peri PL, Pepper DA, Silva LCR, Sheil D, Bonser SP, et al. Carbon isotopic signatures of soil organic matter correlate with leaf area index across woody biomes. Journal of Ecology. 2014;**102**(6):1606-1611. DOI: 10.1111/1365-2745.12309

[66] Walling DE, Zhang Y, He Q. Models for deriving estimates of erosion and deposition rates from fallout radionuclide (Caesium-137, excess Lead-210, and Beryllium-7) measurements and the development of user friendly software for model implementation. In: Impact of Soil Conservation Measures on Erosion Control and Soil Quality. Vienna; IAEA-TECDOC-1665; 2011. pp. 11-33

[67] Conrad O, Bechtel B, Bock M, Dietrich H, Fischer E, Gerlitz L, et al. System for automated geoscientific analyses (SAGA) v. 2.1.4. Geoscientific Model Development. 2015;**8**:1991-2007. DOI: 10.5194/gmd-8-1991-2015

[68] Li X, McCarty GW. Use of principal components for scaling up topographic models to map soil redistribution and soil organic carbon. Journal of Visualized Experiments. 2018;**140**:e58189. DOI: 10.3791/58189

[69] Moriasi DN, Arnold JG, Van Liew MW, Binger RL, Harmel RD, Veith TL. Model evaluation guidelines for systematic quantification of accuracy in watershed simulations. Transactions of the ASABE. 2007;**50**(3):885-900. DOI: 10.13031/2013.23153

[70] Throne M. Southern Iowa agriculture, 1833-1890: The progress from subsistence to commercial corn-belt farming. Agricultural History Society. 1949;**23**:124-130. DOI: 10.2307/3740927

[71] Broder MW, Wagner GH. Microbial colonization and decomposition of corn, wheat, and soybean residue. Soil Science Society of America Journal. 1988;**52**(1): 112-117. DOI: 10.2136/sssaj1988.03615995005200010020x

[72] Li X, McCarty GW. Application of topographic analyses for mapping spatial patterns of soil properties. In: Pepe A, editor. Earth Observation and Geospatial Analyses Geometry. London: IntechOpen; 2019. pp. 1-32. DOI: 10.5772/57353

[73] Dormann CF, Elith J, Bacher S, Buchmann C, Carl G, Carré G, et al. Collinearity: A review of methods to deal with it and a simulation study evaluating their performance. Ecography. 2013;**36**(1):027-046. DOI: 10.1111/j.1600-0587.2012.07348.x

[74] Heckrath G, Djurhuus J, Quine TA, Van Oost K, Govers G, Zhang Y. Tillage erosion and its effect on soil properties and crop yield in Denmark. Journal of Environmental Quality. 2005;**34**:312-324. DOI: 10.1029/2002GB002010

[75] Van Oost K, Govers G, de Alba S, Quine TA. Tillage erosion: A review of controlling factors and implications for soil quality. Progress in Physical Geography. 2006;**30**(4):443-466. DOI: 10.1191/0309133306pp487ra

Chapter 2

# Metamorphic Zircons Applied for Dating East African Tectono-Metamorphic Event in Central Mozambique

*Vicente Albino Manjate*

## Abstract

The term East African is now used to describe the tectonic, magmatic, and metamorphic activity of Neoproterozoic to earliest Paleozoic age. Metamorphic zircon is the most suitable geochronometer for the determination of both protolithic and metamorphic ages due to its high closure temperature. The study area comprises the Mungari and Macossa-Chimoio nappes (Central Mozambique) tectonically juxtaposed to the Archaean Zimbabwe Craton. We use the metamorphic zircon morphology, Th/U ratios, and U-Pb ages to evaluate the Tectono-Metamorphic Event in central Mozambique. Morphologically, the zircon grains are sub-euhedral to euhedral, prismatic, with dark to gray cores, and narrow dark rims. The cores exhibit homogenous domains and oscillatory zoning. On the other hand, the U-Pb zircon data define Th/U ratios of 0.26–0.66 and 0.06–0.11. Finally, the U-Pb zircon analyses define upper intercept age of 1094 ± 36 Ma and lower intercept age of 498 ± 30 Ma. The zircon grains of the Macossa-Chimoio nappe was metamorphically re-homogenized or recrystallized by East African tectono-metamorphic event from relicts of Mesoproterozoic protolith domains. Thrusting and folding are the main East African reworking mechanisms that generated the metamorphic re-homogenization or recrystallization of the Mesoproterozoic magmatic rocks in the Macossa-Chimoio nappe of Central Mozambique.

**Keywords:** metamorphic zircon, protolith, East African, tectono-metamorphic, Macossa-Chimoio nappe, Mungari nappe

## 1. Introduction

Zircon is a fundamental secondary mineral of granitic rocks, very unsusceptible to sedimentary and metamorphic processes [1]. The term 'metamorphic zircon' is used to describe zircon that has formed in rocks under system-wide metamorphic conditions by a range of different processes [2]. According to [2], the main processes include precipitation from the melt during anatectic melting, sub-solidus nucleation and crystallization (blastogenesis) by diffusion of Zr and Si released from metamorphic breakdown reactions of major silicates and accessory phases,

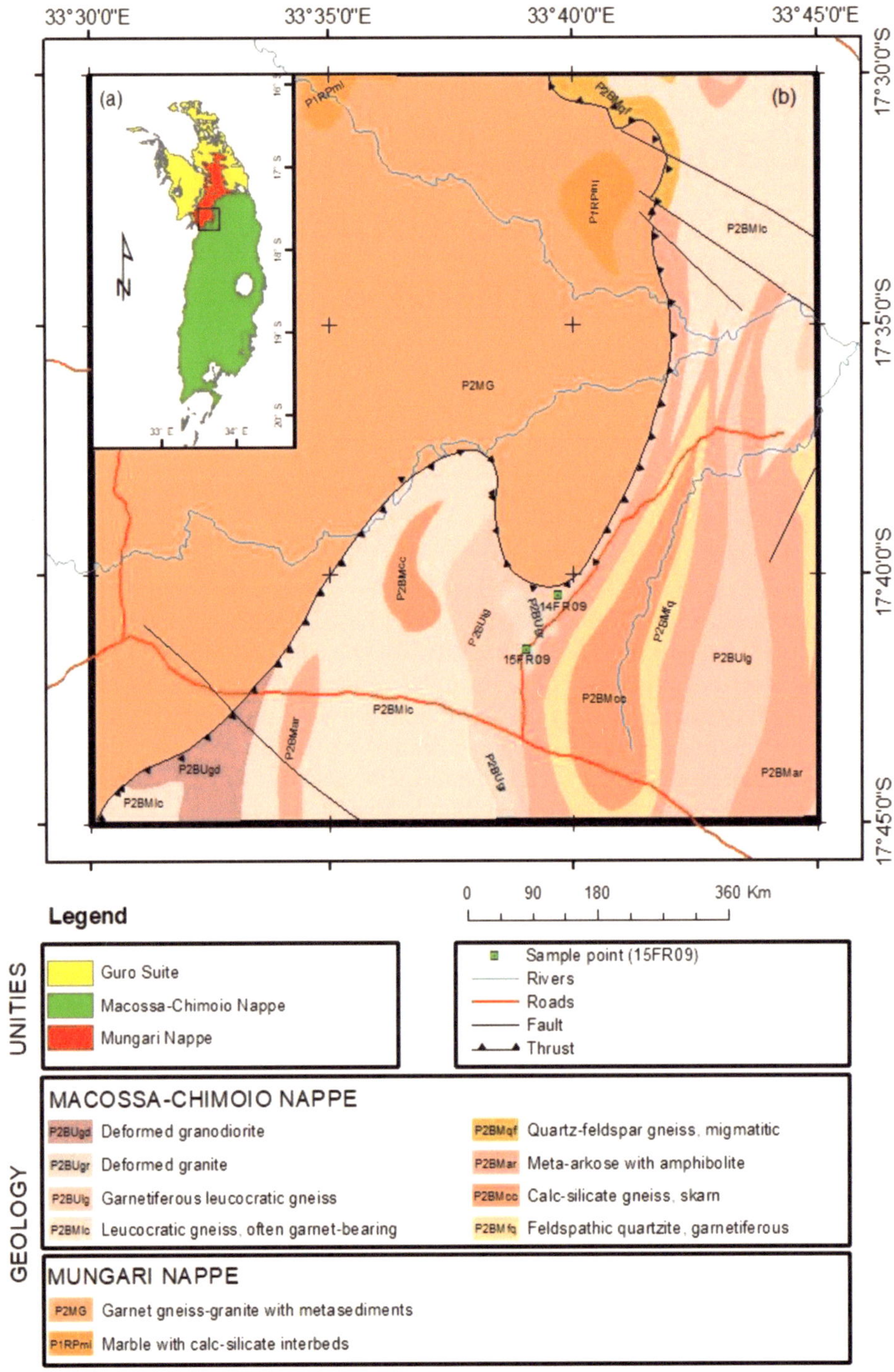


**Figure 1.**
*Geologic setting of the study area. (a) Regional geologic unities and (b) geology of the Macossa-Chimoio and Mungari nappes. Modified from [5–8].*

precipitation from aqueous metamorphic fluid, and protolith zircon recrystallization. For these authors, knowing which process is responsible for the genesis of 'metamorphic zircon' in a particular sample is crucial for the correct interpretation of U-Pb isotopic data and derived ages, and consequently the interpretations of whole-rock petrogenesis.

| Method | LA-ICP-MS | SHRIMP | TIMS |
|---|---|---|---|
| Applicability | • U-Pb zircon | • U-Pb zircon, titanite | • U-Pb zircon |
| Advantage | • Relatively cheap;<br>• Very quick (~2 min per analysis);<br>• Fairly precise with an internal error of ~1%;<br>• Spot size = 29 μm;<br>• Excellent sensitivity, precision and good accuracy for isotope ratio measurements | • Accurate with an external error of ~1%;<br>• fast with the time of analyses ~10–15 min;<br>• Primary beam analytical spot size = 30 μm | • Ultrahigh precision in U-Pb dating |
| Disadvantage | • Does not always produce consistent results within error | • High cost partly limits its wide application | • Requires ultraclean laboratory;<br>• The sample preparation is time-consuming |

*Modified from [11, 14, 27].*

**Table 1.**
*Summary of the applicability, advantage, and disadvantage of the LA-ICP_MS, SHRIMP, and TIMS dating techniques.*

The term East African is applied to illustrate the tectonism, magmatism, and metamorphism that took place on Neoproterozoic to earliest Paleozoic, mainly for a crust that was formally portion of Gondwana [3]. The term 'East African' was suggested by [4] supported on isotopic ages of Africa by Rb-Sr and K-Ar methods. According to [3], the East African was explained as a Neoproterozoic tectono-thermal event (~500 Ma) during which a number of mobile belts produced, bounding older cratons. This tectono-thermal event constitutes the final stage of an orogenic cycle, conducting to orogenic belts presently interpreted as a consequence of the fusion of continental blocks throughout the time interval from ~870 to ~550 Ma [3].

The study area comprises the nappes of Macossa-Chimoio and Mungari [5] (**Figure 1**). According to [5], the northern Mungari nappe is composed of meta-sedimentary supracrustal rocks intruded by a set of granitoid plutons. On the other hand, the southern Macossa-Chimoio nappe is composed of orthomagmatic rocks covered by medium to high-grade supracrustal rocks. In addition, both nappe complexes include detrital zircon grains with Neoproterozoic age [6].

We use the metamorphic zircon morphology, Th/U ratios, and $^{207}Pb/^{206}Pb$ ages to evaluate the East African tectono-metamorphic event in Central Mozambique. Although the SHRIMP technique is very expensive, its advantages in comparison to other dating techniques are in favour of the geochronological data determination for this study (**Table 1**). One of the most important legacies of SHRIMP U-Pb dating on zircons is the extraction of crystallization and recrystallization ages of igneous protoliths from complexly deformed and metamorphosed lithologies.

## 2. Geological and tectonic setting

The study area (**Figure 1**) comprises rocks of the Neoproterozoic Mungari and the Mesoproterozoic Macossa-Chimoio nappes tectonically juxtaposed to the Archaean Zimbabwe Craton [5–7]. The Neoproterozoic Mungari nappe is composed of garnet gneiss-granite plutons of about 850 Ma intruding meta-sedimentary rocks consisting of marbles with calc-silicate interbeds [7]. This nappe is delimited on the west, north and east by Neoproterozoic (~850 Ma) bimodal Guro Suite and on the south by the Macossa-Chimoio nappe [7, 8]. The Mesoproterozoic Macossa-Chimoio nappe consists of medium- to high-grade supracrustal rocks composed of quartz-feldspar gneiss, deformed granodiorite, deformed granite, garnetiferous leucocratic gneiss, meta-arkose with amphibolite, calc-silicate gneiss, feldspathic quartzite, and leucocratic gneiss [7, 8]. According to [8], in the north, the Macossa-Chimoio nappe terminates into a northward-directed arcuate thrust (**Figure 1**). For [8], the Macossa-Chimoio nappe consists of supracrustal rocks most likely derived from sedimentary precursors, originally deposited in a shallow marine paleobasin. Although the definitive character and position of all units observed within the supracrustal rocks succession are not fully solved, the overall lithostratigraphy of the paleobasin has been reduced by [8] from several geological sections made in the area [9, 10]. The lowermost rock units of the inferred paleobasin include garnetiferous leucocratic gneisses, quartz-feldspar gneisses, meta-arkoses, and arkosic quartzites. These psammitic metasediments are overlain by more pelitic rocks (metagreywackes, garnet, and sillimanite bearing mica schist and mica gneisses) with thin calc-silicate gneiss and marble interbeds.

The Macossa-Chimoio nappe is delimited by a number of structural domains. According to [8], the eastern margin of the Mesoproterozoic Macossa-Chimoio nappe is bounded by a set of rift faults/dykes 'corridor' against the Karoo and younger formations and partly remains covered by recent sediments, the western margin is a major N-S directed sinistral shear zone along the Archaean cratonic margin, in the north the nappe terminates into a northward-directed thrust, and in the south the rocks of the nappe become covered by Phanerozoic beds. According to [8], the northward thrusting of the northern part of the Macossa-Chimoio nappe over the Mungári nappe gneisses may be attributed to the East African collision, the sinistral shearing is a regional feature in the East Africa orogeny, and the set of rift faults/dykes 'corridor' against the Karoo and younger formations are normal faults with dip values commonly ranging from 45 to 60°.

The granitic pluton (deformed granite) selected for this study is located at the northern end of the Macossa-Chimoio nappe and was emplaced parallel to the foliation of the host gneisses and migmatites. The rock is a pinkish to pinkish gray, medium- to coarse-grained, weakly deformed leucogranite and is mainly composed of quartz, pinkish potassium feldspar, plagioclase, hornblende, and biotite. Accessory minerals include garnet, clinopyroxene, orthopyroxene, zircon, apatite, and opaques.

## 3. Analytical procedures

Zircon dating analyses by sensitive high-resolution ion microprobe (SHRIMP) U-Pb were performed at the São Paulo University, Brazil. This technique is very important for geochronological studies [11] as it permits *in situ* analyses of complex zircons grains often exhibiting several crystallization phases associated with different geological processes [12]. According to [12], the SHRIMP technique has an improving spatial resolution for dating with precision the different growth episodes on single zircon grains. Zircon crystals were separated utilizing the common manually breaking, crushing and grinding of samples, followed by grain size separation by sieving. The material (100–200 mesh portion) was deposited on a vibrating Wilfley table, and heavy minerals were then densimetrically separated using bromoform (d = 2.89 g/ml; 20°C) and methylene iodide (d = 3.32 g/ml; 20°C). The dense material (density above 3.32 g/ml) was then electromagnetically separated using a Frantz separator. The magnetic minerals were separated using a hand magnet and the paramagnetic minerals were separated by a Frantz

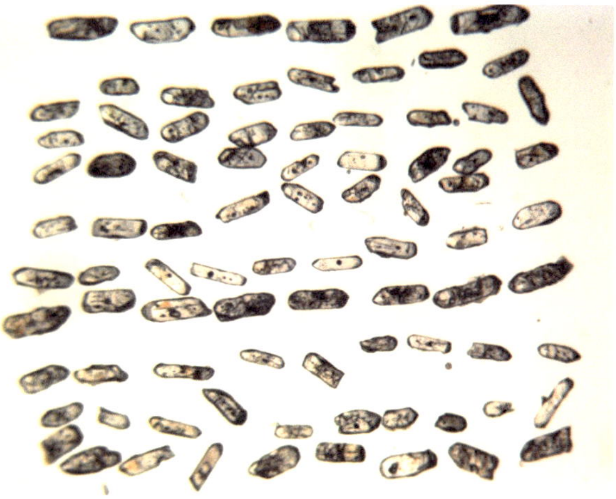

**Figure 2.**
*Reflected and transmitted light images of the deformed granite zircons. Photograph length of 2.3 mm.*

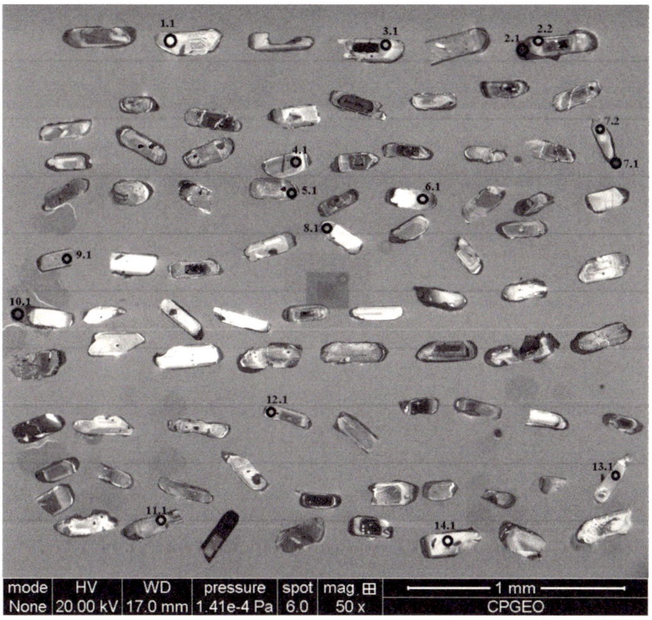


**Figure 3.**
*CL image of zircons with selected analyses locations (spots) by SHRIMP IIe/MC.*

magnetic separator (amperage variation). Zircon crystals are concentrated in the non-magnetic portion. Zircon crystals along with zircon standards were picked by hand, impregnated in epoxy resin mounts with a diameter of 2.54 cm, ground and polished with diamond compound (1–7 μm) to reveal grain centers and carbon coated as well as cleaned and gold-coated in preparation for the SHRIMP analyses. Reflected and transmitted light images (**Figure 2**) were acquired before the gold coating of 2–3 μm. Zircons internal structures were microphotographed in transmitted and reflected light and characterized by the use of cathodoluminescence (CL) images from scanning electron microscope prior to SHRIMP U-Pb zircon isotopic analyses. CL images of representative zircon crystals can be seen in **Figure 3**. After CL acquisition, the gold was removed and the mount was re-cleaned. The U-Pb zircon dating analyses were made using a SHRIMP IIe/MC mass spectrometer

| Spot | $^{206}Pbc$ (%) | U (ppm) | Th (ppm) | $^{232}Th/^{238}U$ | $^{206}Pb^*$ (ppm) | Ages (Ma) corrected to $^{204}Pb$ | | | | Disc (%) | Ratios corrected to $^{204}Pb$ | | | | Error correl |
|---|---|---|---|---|---|---|---|---|---|---|---|---|---|---|---|
| | | | | | | $^{206}P/^{238}U$ | Error (%) | $^{207}Pb/^{206}Pb$ | Error (%) | | $^{207}Pb^*/^{235}U$ | Error (%) | $^{206}Pb^*/^{238}U$ | Error (%) | |
| Amostra 15FR09 | | | | | | | | | | | | | | | |
| **1.1c** | 0.19 | 127 | 55 | 0.44 | 21.6 | 1159.5 | 15.0 | 1102 | 77 | −5 | 2.07 | 4.1 | .1971 | 1.4 | .346 |
| **2.1r** | 0.01 | 890 | 88 | 0.10 | 61.3 | 497.0 | 5.8 | 542 | 16 | 9 | 0.64 | 1.4 | .0801 | 1.2 | .861 |
| **2.2c** | 0.38 | 275 | 69 | 0.26 | 31.3 | 799.9 | 15.9 | 875 | 43 | 9 | 1.24 | 3.0 | .1321 | 2.1 | .714 |
| **3.1c** | 0.06 | 175 | 74 | 0.43 | 27.7 | 1090.8 | 13.0 | 1136 | 19 | 4 | 1.97 | 1.6 | .1844 | 1.3 | .804 |
| **4.1c** | −0.05 | 210 | 100 | 0.49 | 30.5 | 1006.4 | 12.0 | 1053 | 30 | 5 | 1.73 | 2.0 | .1690 | 1.3 | .656 |
| **5.1c** | 0.14 | 472 | 133 | 0.29 | 65.5 | 963.1 | 11.0 | 971 | 23 | 1 | 1.59 | 1.7 | .1611 | 1.2 | .740 |
| **6.1c** | 0.05 | 206 | 94 | 0.47 | 27.6 | 933.4 | 11.8 | 1030 | 27 | 10 | 1.58 | 1.9 | .1558 | 1.4 | .708 |
| **7.1r** | 0.39 | 942 | 56 | 0.06 | 69.0 | 525.8 | 6.2 | 515 | 38 | −2 | 0.67 | 2.1 | .0850 | 1.2 | .580 |
| **7.2c** | 0.07 | 243 | 135 | 0.57 | 39.9 | 1127.8 | 13.1 | 1114 | 15 | −1 | 2.02 | 1.5 | .1912 | 1.3 | .856 |
| **8.1c** | 0.06 | 279 | 98 | 0.36 | 42.6 | 1056.0 | 13.9 | 1136 | 16 | 8 | 1.90 | 1.6 | .1780 | 1.4 | .871 |
| **9.1c** | 0.05 | 275 | 165 | 0.62 | 39.7 | 1000.6 | 11.6 | 1057 | 14 | 6 | 1.73 | 1.4 | .1679 | 1.2 | .871 |
| **10.1r** | 0.20 | 860 | 95 | 0.11 | 59.4 | 497.7 | 5.8 | 510 | 23 | 2 | 0.64 | 1.6 | .0803 | 1.2 | .755 |
| **11.1c** | 0.29 | 238 | 97 | 0.42 | 26.5 | 782.2 | 9.5 | 872 | 31 | 11 | 1.21 | 2.0 | .1290 | 1.3 | .650 |
| **12.1c** | 0.09 | 288 | 111 | 0.40 | 44.2 | 1060.1 | 12.5 | 1115 | 16 | 5 | 1.89 | 1.5 | .1787 | 1.3 | .847 |
| **13.1c** | 0.15 | 250 | 145 | 0.60 | 34.4 | 957.5 | 11.3 | 1025 | 23 | 7 | 1.62 | 1.7 | .1601 | 1.3 | .751 |
| **14.1c** | 0.32 | 110 | 70 | 0.66 | 17.1 | 1066.0 | 13.2 | 1032 | 35 | −3 | 1.83 | 2.2 | .1798 | 1.3 | .610 |

*Total radiogenic.*

**Table 2.**
*Analytical data for zircons from the deformed granite, Macossa-Chimoio nappe.*

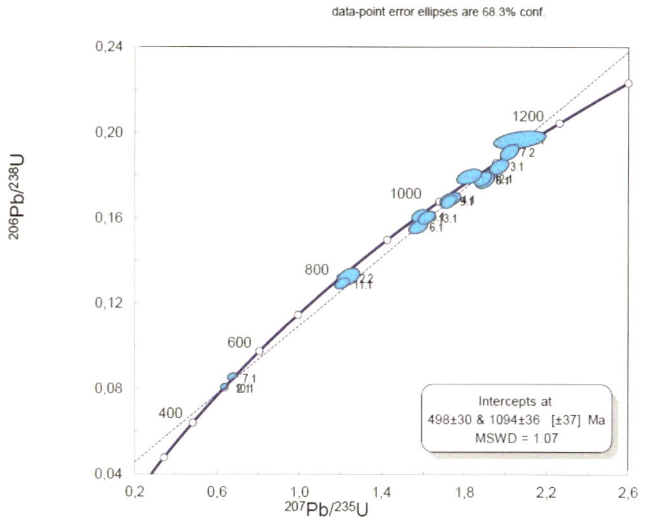


**Figure 4.**
*Concordia diagram of zircon U-Pb isotope data analyzed by SHRIMP IIe/MC for the deformed granite, Macossa-Chimoio nappe.*

and zircon standards designated Temora 2 [13]. As described in [14], this consisted of measuring U, Pb and Th abundances, and isotopic relationships of these elements in zircon crystals. The precision of the U, Pb and Th zircons analytical data obtained by SHRIMP IIe at São Paulo University Laboratory is of the same standard when compared to the data from leading laboratories worldwide [14]. The reduction of raw data was carried out using SQUID 1.06 [15]. Common lead corrections usually use $^{204}Pb$ according to [16].

The U-Pb zircon dating results are shown in **Table 2** and projected in the Concordia diagram (see **Figure 4**). There is evidence of recrystallized domains and inherited cores on CL images. Results from recrystallized domains and inherited cores were not used in the final age determination. Results above 10% discordance (1σ) and/or with extremely big errors originated by a correction of common lead were also not used in the final age determination. Age determination and Concordia diagram processed using ISOPLOT of version 4.0 [15]. Errors shown in **Table 2** and **Figure 4** are 1σ levels. In a situation of evident Pb-loss, consecutive younger results were discarded until obtaining an acceptable mean square of weighted deviations (MSWD). The remaining U-Pb zircon dating results were used to determine the magmatic and metamorphic ages. All age errors in the text and Concordia diagram are considered at (1σ).

## 4. Results and discussions

### 4.1 Zircon morphology and internal structure

Zircon morphology and internal structure provide an important tool for discerning their growth stages and genesis. Zircons of our study are products of anatectic melt, altered by metamorphic fluid and hydrothermal alteration. Anatectic melt, metamorphic fluid, and hydrothermal alteration are important factors controlling the morphology and internal structure of zircons overgrowths [17]. Zircons crystallizing from anatectic melts also have a euhedral shape with no zoning, planar zoning or oscillatory zoning. In addition, zircons altered by the metamorphic fluid are usually homogenous with high CL intensity, showing resorption structure. Moreover, zircon domains that lost all radioactive Pb during hydrothermal alteration always show white color in CL image. The study made on deformed granite showed that the zircon grains are inherited from older crustal rocks or metamorphically re-homogenized or recrystallized from relicts of magmatic protolith domains. These zircon grains range in size from 250 to 400 μm of length, and 125 μm of width with length/width ratios of 2:1–3.2:1. In addition, they are colorless and transparent (**Figure 2**), sub-euhedral to euhedral with elongated prismatic shapes. Moreover, these zircon grains exhibit narrow dark rims and dark to gray cores with some homogenous domains and other domains of compositional or oscillatory zoning (**Figure 3**), being thus strongly re-homogenized. Therefore, the zircons of our study are products of anatectic melts or altered by metamorphic and hydrothermal fluids.

### 4.2 Genesis and recrystallization of the metamorphic zircons

Sixteen analyses spots (cores and rims) in 14 zircon grains from the deformed granite for Th/U ratios and $^{207}Pb/^{206}Pb$ ages determinations (**Table 2**). These analyses spots define two groups based on Th/U ratios and apparent $^{207}Pb/^{206}Pb$ ages. The first group is defined by cores with U grades from 110 to 472 ppm (averaging 242 ppm) and Th ranging from 55 to 165 ppm (averaging 104 ppm). This result in Th/U ratios from 0.26 to 0.66 and $^{207}Pb/^{206}Pb$ ages varying from 872 ± 31 to 1136 ± 19 Ma. The other group is represented by rims with U grades ranging from 860 to 942 ppm (average of 890 ppm) and Th ranging from 56 to 95 ppm (average of 80 ppm). This result in Th/U ratios from 0.06 to 0.11 (average of 0.09) and $^{207}Pb/^{206}Pb$ ages varying from 510 ± 23 to 542 ± 16 Ma. The age data follow a regression line (**Figure 4**) that allowed to determine the upper intercept age of 1094 ± 36 Ma and lower intercept age of 498 ± 30 Ma (MSWD = 1.07).

Th/U ratios are used as indicators of zircon types. The Th/U ratios of magmatic zircons are commonly between 0.32 and 0.70, whereas hydrothermal zircons frequently have more extreme values [18–20]. Proposed that Th/U ratios <0.1 are probably a hint for hydrothermal origin. Therefore, the studied zircons are products of metamorphic re-homogenization or recrystallization from relicts of magmatic protoliths.

The studied metamorphic zircons registered two events. The magmatic protolith domains crystalized at ~1094 ± 36 Ma. This was followed by re-homogenization or recrystallization related to northward-directed thrusting and folding at ~498 ± 30 Ma of the Chimoio-Macossa nappe [5]. Using LA-ICP-MS U/Pb zircon for leucocratic gneiss (sample 14FR09, **Figure 1**) of Chimoio-Macossa nappe found $^{207}Pb/^{206}Pb$ crystallization age of 1067.8 ± 9.0 Ma and a metamorphic age of 504 ± 1.8 Ma. These age determinations are in accordance with that of [21] for the Macossa-Chimoio nappe. According to Yuanbao (2004), the time of metamorphic recrystallization is represented by the age of recrystallized zircon domain with the lowest Th/U ratio and the youngest U-Pb age.

The Cambrian U-Pb ages are found in both cores and rims of the Mesoproterozoic Macossa-Chimoio nappe rocks. Manjate [22] found a Neoproterozoic-Cambrian recrystallization age (498 ± 19–562 ± 14 Ma) on zircon cores of the Dongueni Mount nepheline syenite generated from partial melting of Mesoproterozoic crust, as shown by inherited zircon ages of 1040 ± 14 (15) Ma. The Cambrian magmatism, defined by zircon U-Pb dates of c. 490 Ma, from Dongueni Mount nepheline syenite, southeast of Chimoio village [23], is post-collisional and marks the end stage of East African Orogeny. Therefore, zircons are suitable for dating the tectono-metamorphic Neoproterozoic-Cambrian event that affected the Macossa-Chimoio nappe. The Neoproterozoic-Cambrian recrystallization ages on zircon were the common determinations made by a number of authors [23–26].

## 5. Conclusions

Zircons are appropriate for dating the tectono-metamorphic Neoproterozoic-Cambrian event that affected the Macossa-Chimoio nappe. The studied zircon grains exhibit narrow dark rims and dark to gray cores with some homogenous domains and other domains of compositional or oscillatory zoning, as well as, Th/U ratios less than 0.3 which are evidence of metamorphic re-homogenization or recrystallization from relicts of magmatic protolith domains.

Thrusting and folding are the main East African Neoproterozoic-Cambrian reworking mechanisms (ca. 498 ± 30 Ma) that generated the metamorphic re-homogenization or recrystallization of the Mesoproterozoic magmatic rocks (ca. 1094 ± 36 Ma) in the Macossa-Chimoio nappe of Central Mozambique.

## Acknowledgements

This study results from the PhD project involving the cooperation of institutions (CNPQ and PRO-AFRICA, Brazil, and the National Institute of Mines, Mozambique) in the form of financial support and services. The author thanks Centro de Pesquisas Geocronologicas (CPGeo) of IGc/USP for the U-Pb zircon analyses and anonymous reviewers for their careful comments that have improved the manuscript remarkably.

## Conflict of interest

The author declares that there is no conflict of interest.

## Author details

Vicente Albino Manjate
National Institute of Mines, Maputo, Mozambique

*Address all correspondence to: vmanjate@yahoo.com.br

## References

[1] Martins HCB, Simões PP, Abreu J. Zircon crystal morphology and internal structures as a tool for constraining magma sources: Examples from northern Portugal Variscan biotite-rich granite plutons. Comptes Rendus Geoscience. 2014;**346**:233-243

[2] Hoskin PWO, Black LP. Metamorphic zircon formation by solid-state recrystallization of protolith igneous zircon. Journal of Metamorphic Geology. 2000;**18**:423-439

[3] Kröner A, Stern RJ. Pan-African Orogeny. Africa. Encyclopedia of Geology. Vol. 1. Amsterdam: Elsevier; 2004. pp. 1-12

[4] Stern RJ. ARC assembly and continental collision in the neoproterozoic east african orogen: Implications for the consolidation of gondwanaland. Annual Review of Earth and Planetary Sciences. 1994;**22**:319-351. DOI: 10.1146/annurev.ea.22.050194.001535

[5] Chaúque FR, Cordani UG, Jamal DL. Geochronological systematics for the Chimoio-macossa frontal nappe in Central Mozambique-implications for the tectonic evolution of the southern part of the Mozambique belt. Journal of the African Earth Sciences. 2019;**150**:47-67. DOI: 10.1016/j.jafrearsci.2018.10.013

[6] Chaúque FR, Cordani UG, Jamal DL, Onoe AT. The Zimbabwe craton in Mozambique: A brief review of its geochronological pattern and its relation to the Mozambique belt. Journal of the African Earth Sciences. 2017;**129**:366-379. DOI: 10.1016/J.JAFREARSCI.2017.01.021

[7] Chaúque FR. Contribuição para o conhecimento da evolução tectônica do Cinturão de Moçambique, em Moçambique. Biblioteca Digital de Teses e Dissertações da Universidade de São Paulo; 2012. DOI: 10.11606/T.44.2012.tde-02062015-152355

[8] GTK Volume II. Map explanation: Sheets 1630-1934. Geology of Degree Sheets Mecumbura, Chioco, Tete, Tambara, Guro, Chemba, Manica, Catandica, Gorongosa, Rotanda, Chimoio and Beira; Maputo:vol. 2; 2006

[9] Hunting. Ground geophysics. mineral inventory project in tete province and parts of manica, sofala and zambezia provinces. Report on ground geophysics investigations for the period July to October 1982; Maputo: 1983

[10] Hunting. Mineral inventory project in tete province and parts of manica, sofala and zambezia provinces. Report on ground geophysical investigations for the 1982 and 1983 field season; Maputo: 1984

[11] Sato K, Junior OS, MAS B, CCG T, Onoe AT. SHRIMP U-Th-Pb analyses of titanites: Analytical techniques and examples of terranes of the south-southeast of Brazil: Geoscience Institute of the University of São Paulo. Geologia USP. Série Científica. 2016;**16**:3-18. DOI: 10.11606/issn.2316-9095.v16i2p3-18

[12] Chemale F Jr, Kawashita K, Dussin IA, Ávila JN, Justino D, Bertotti A. U-Pb zircon in situ dating with LA-MC-ICP-MS using a mixed detector configuration. Annals of the Brazilian Academy of Sciences. 2012;**84**:275-295

[13] Black LP, Kamo SL, Allen CM, Davis DW, Aleinikoff JN, Valley JW, et al. Improved 206Pb/238U microprobe geochronology by the monitoring of a trace-element-related matrix effect; SHRIMP, ID–TIMS, ELA–ICP–MS and oxygen isotope documentation for a series of zircon standards. Chemical Geology. 2004;**205**:115-140. DOI: 10.1016/j.chemgeo.2004.01.003

[14] Sato K, Tassinari CCG, Basei MAS, Siga Júnior O, Onoe AT, de

Souza MD. Sensitive high resolution ion microprobe (SHRIMP IIe/MC) of the institute of geosciences of the University of São Paulo, Brazil: Analytical method and first results. Geologia USP. Série Científica. 2014;**14**:3-18. DOI: 10.5327/Z1519-874X201400030001

[15] Ludwig KR. User's Manual for Isoplot 3.00: A Geochronological Toolkit for Microsoft Excel. Berkeley CA: Special publication/Berkeley Geochronology Center; 2003

[16] Stacey JS, Kramers JD. Approximation of terrestrial lead isotope evolution by a two-stage model. Earth and Planetary Science Letters. 1975;**26**:207-221. DOI: 10.1016/0012-821X(75)90088-6

[17] Yuambao W, Yongfei Z. Genesis of zircon and its constraints on the interpretation of U-Pb age. Chinese Science Bulletin. 2004;**49**:1554. DOI: 10.1360/04wd0130

[18] Fu B, Mernagh TP, Kita NT, Kemp AIS, Valley JW. Distinguishing magmatic zircon from hydrothermal zircon: A case study from the Gidginbung high-sulphidation Au–Ag–(Cu) deposit, SE Australia. Chemical Geology. 2009;**259**:131-142. DOI: 10.1016/J.CHEMGEO.2008.10.035

[19] Rubatto D. Zircon trace element geochemistry: Partitioning with garnet and the link between U-Pb ages and metamorphism. Chemical Geology. 2002;**184**:123-138. DOI: 10.1016/S0009-2541(01)00355-2

[20] Hoskin PWO, Schaltegger U. The composition of zircon and igneous and metamorphic petrogenesis. Reviews in Mineralogy and Geochemistry. 2003;**53**:27-62. DOI: 10.2113/0530027.

[21] Fritz H, Abdelsalam M, Ali KA, Bingen B, Collins AS, Fowler AR, et al. Orogen styles in the east African orogen: A review of the Neoproterozoic to Cambrian tectonic evolution. Journal of the African Earth Sciences. 2013;**86**:65-106. DOI: 10.1016/j.jafrearsci.2013.06.004

[22] Manjate VA. Whole-rock geochemical, U-Pb and Sm-Nd isotope characteristics of the Dongueni Mont nepheline syenite intrusion, Mozambique. Geoscience Frontiers. 2015;**8**:1063-1071. DOI: 10.1016/j.gsf.2016.10.009

[23] Manjate VA. Geocronologia da região de Gondola-Nhamatanda (Centro de Moçambique). Biblioteca Digital de Teses e Dissertações da Universidade de São Paulo; 2012. DOI: 10.11606/D.44.2012.tde-21122012-085416

[24] Manjate VA. Caracterização geocronológica dos granitóides do complexo de bárue e da suíte de guro, centro-oeste de moçambique: Iplicações tectônicas e metalogenéticas. Biblioteca Digital de Teses e Dissertações da Universidade de São Paulo; 2015. DOI: 10.11606/T.44.2015.tde-22122015-143300

[25] Manjate VA. U-Pb zircon geochronology and Sr-Nd isotopic composition of the Inchope orthogneiss in Mozambique: Age constraints and petrogenetic implications. Journal of the African Earth Sciences. 2017;**131**:98-104. DOI: 10.1016/j.jafrearsci.2017.03.027

[26] Manjate VA, Tassinari CCG. Zircon U-Pb geochronology and Nd isotope systematics of the Guro suite granitoids, Mozambique: Implications for Neoproterozoic crust reworking events. Journal of the African Earth Sciences. 2018;**148**:69-79. DOI: 10.1016/j.jafrearsci.2018.05.012

[27] Becker JS. State-of-the-art and progress in precise and accurate isotope ratio measurements by ICP-MS and LA-ICP-MS: Plenary lecture. Journal of Analytical Atomic Spectrometry. 2002;**17**:1172-1185. DOI: 10.1039/B203028B

Section 2

# Applications in Hydrosphere

Chapter 3

# Isotopic Application in High Saline Conditions

*Minghui Li, Xiaomin Fang, Jiao Li, Maodu Yan, Shurui Sun and Liping Zhu*


## Abstract

Evaporite minerals record the hydrogeochemical conditions in which they precipitated. And therefore they can be used to reconstruct the paleoclimate and paleoenvironments. Evaporite minerals are also major sources of industrial minerals including gypsum, halite, borates, lithium concentrates, and others. Because of their scientific significance and economic importance, evaporite minerals and their isotopic hydrochemical processes linked to their formation have been the focus of many geologists and paleoclimatologists. This chapter will discuss the application of isotopes of hydrogen, oxygen, sulfur, strontium, and boron in saline conditions. This will include the following: the $\delta^{18}O$ and $\delta D$ of hydrated water of gypsum and their paleoclimate since 2.2 Ma in the Qaidam Basin, NE Tibetan Plateau; the $\delta^{18}O$ and $\delta D$ of the interlayer water of clay minerals in salar lacustrine sediments; and the $^{87}Sr/^{86}Sr$, $\delta^{34}S$, and $\delta^{11}B$ of halite from evaporite deposits in Khorat Plateau, Laos, and Yunnan and their application in the origins of brine.


**Keywords:** oxygen and hydrogen, Sr isotope, sulfur isotope, boron isotope, salt minerals

## 1. Introduction

Evaporite minerals record the hydrogeochemical conditions in which they precipitated. And therefore they can be used to reconstruct the paleoclimate and paleoenvironments. Evaporite minerals are also major sources of industrial minerals including gypsum, halite, borates, lithium concentrates, and others. Given their scientific significance and economic importance, evaporite minerals and their isotopic hydrochemical processes linked to their formation have become the focus of many geologists and paleoclimatologists. The Qaidam Basin and Lanping-Simao Basin (LSB) in China and Khorat Basin (KB) and Sakhon Nakhon Basin (SNB) in Thailand and Laos, which have thick sequences of evaporites, have been the focus of study by researcher concerned with its mineral resources, geochemistry, paleoenvironmental evolution, and tectonic uplift [1–3].

Water is nearly ubiquitously in nature. Its isotopic compositions also were recorded in many minerals. In order to identify the past conditions at which sediments formed, many studies focused on the stable isotopes of carbon and oxygen as tracers of environmental conditions. However, we know little about the $\delta^{18}O$ and $\delta D$ in saline conditions. On the other hand, halite and gypsum are two common minerals in evaporite deposits. Today, exploited evaporite deposits are most found in the arid and semiarid deserts of the world (the areas between latitudes 15° and 45° both north and south of the equator). Tectonics, climate, and origins of brine are the prime controls on most evaporite deposits. However, the origins of brine was a big topic for some giant evaporite deposits, for example, the evaporite deposit in Khorat Basin (KB) in Thailand, Sakhon Nakhon Basin (SNB) in Laos, and Lanping-Simao Basin (LSB) in China. Almost all large potash deposits are associated with marine fluids (such as the deposits in Thailand, the United States, Germany, Russia, France, and Brazil), whereas some small potash deposits are with fluids of continental source (such as deposits in the Qaidam Basin, Western Tibet, China) [1, 2]. The isotopes of S, Sr, and B of halite could be very useful in determining the origins of brine. So this chapter will use well-known accepted methods to analyze minerals (halite, gypsum, and clay minerals) and stable isotopes ($\delta^{18}O$ and $\delta D$, $^{87}Sr/^{86}Sr$, $\delta^{34}S$, and $\delta^{11}B$) to discuss their application in paleoclimate and origins of brine.

## 2. The $\delta^{18}O$ and $\delta D$ in saline conditions

### 2.1 Geological setting and the Core SG-1 in the Qaidam Basin, NE Tibetan Plateau

The Qaidam Basin is a Mesozoic-Cenozoic sedimentary basin located on the northern margin of the Tibetan Plateau in China (**Figure 1**). The Asian inland drought basin was formed as the result of intracontinental deformation and plateau uplift due to the collision of the Indian and Eurasian Plates. It is bordered by the Kunlun Mountains to the south, the Qilian Mountains to the northeast, and the Arjin Mountains to the northwest, which have altitudes ranging from 4000 to 4500 m asl to over 5000 m asl. A network of faults exists within the basin [3]. The Paleogene and Neogene strata are widespread and include intrusive rocks. From the Middle Oligocene to the Upper Pliocene, the strata consist of mudstones, calcareous mudstones and marls, intercalating siltstone, very thick gypsum, and rock salt beds in the western basin [4]. Some secondary or subsidiary basins have become well developed since the Middle Pleistocene as a result of slow uplift [1]. Lacustrine sediments in

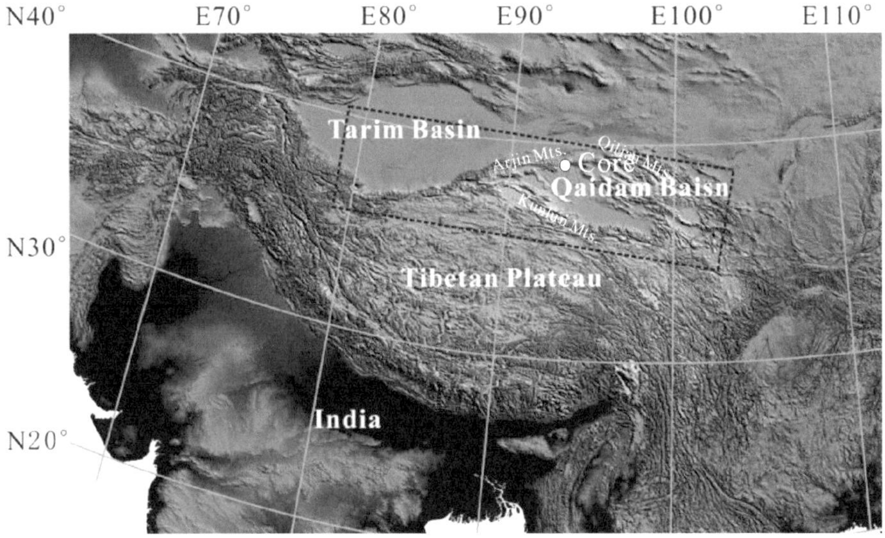


**Figure 1.**
*Map of the Qaidam Basin showing the core site.*

the basin cover an area of 30,000 km$^2$ with strata up to 700 m thick [4]. Abundant records of palaeo-lake level and paleoclimatic changes are indicated by various proxies, such as pollen, ostracods, evaporite minerals, and isotopic geochemistry [5, 6]. Studies on evaporite minerals and hydrochemistry and geochemistry of salt lakes have been ongoing for about 50 years because of great economic significance [1, 2, 5, 7]. However, no two drainage basins have identical climatic and hydrologic conditions, and even adjacent basins can show striking variations in evaporite mineralogy [8].

The Core SG-1 (38°24'35.3" N, 92°30'32.7" E, 2900 m asl) was located in a playa in the Qahansilatu sub-basin, western Qaidam Basin (**Figure 1**). The core spanned from 2.77 to 0.1 Ma, dated by paleomagnetism and optically stimulated luminescence (OSL) with a sedimentary rate of 26.1–51.5 cm/ka [9]. The sedimentary sequence is composed of clay, clay-silt, and siltstone intercalated with salt layers (mainly halite), marl beds, and thin and/or scattered gypsum crystals, indicating clay-silt and halite-marl depositional cycles (**Figure 2**). The fluctuation between evaporite minerals and carbonaceous clay strata indicates a shift between dry and wet climates. Gypsum and halite make up the majority of the evaporite minerals in the core [11, 12]. During the last ~2.8 Ma, the lake basin evolved from a semi-deep fresh lake to a semi-brackish lake (2.8–2.2 Ma), to a perennial saline lake (2.2–2.0 Ma), to a shallow brackish lake (1.8–1.6 Ma), to a perennial saline lake (1.2–0.6 Ma), to a playa saline lake (0.9–0.6 Ma), and to a saline mudflat (0.6–0.1 Ma) [10] (**Figure 2**). Multi-proxies, such as sedimentary features [10], salt minerals [11], isotope records [13, 14], and rare earth elements [15] of the study area, indicate the long-term persistent aridification of inland Asia since 2.8 Ma.

## 2.2 Clay minerals and the $\delta^{18}O$ and δD of their interlayer water

### 2.2.1 Clay minerals

Clay minerals have been used to reconstruct paleoclimates and environments since Singer's review in 1984 [16]. It is common that the clay minerals undergo

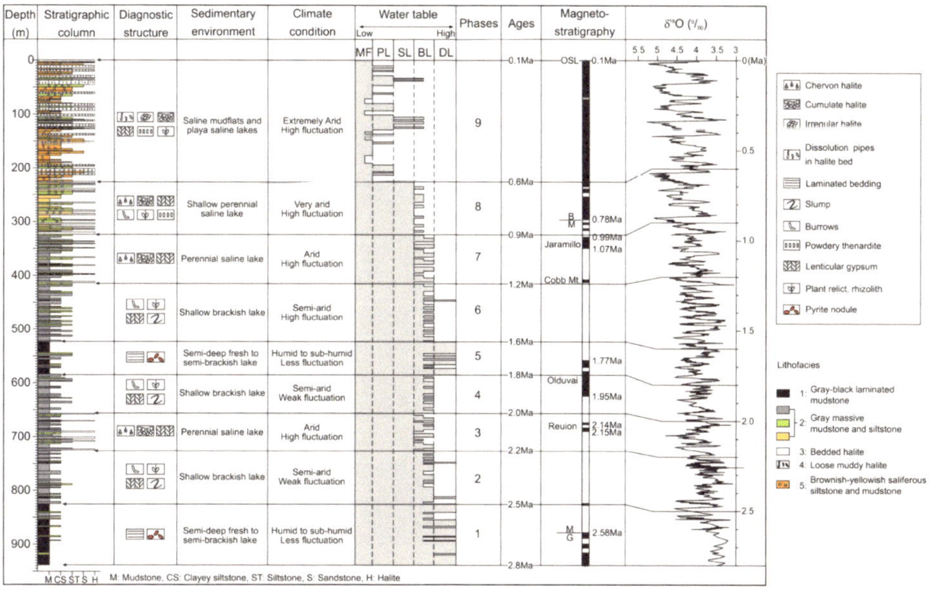


**Figure 2.**
*Lithology, sedimentary environment, sediment accumulation rates (SAR), and magnetostratigraphic ages of the Core SG-1 (after Zhang et al. [9] and Wang et al. [10]). DL, semi-deep fresh to semi-brackish lake; BL, shallow brackish lake; SL, perennial saline lake; PL, playa saline lake; and MF, playa saline mudflat.*

diagenetic transformations or postdepositional diagenetic changes, and as a result, some paleoclimate information of clays will be overprinted or changed. However, there are still lots of studies to use clay minerals to reconstruct paleoenvironments and paleoclimates. The isotopic exchange rates between the interlayer water of clay minerals and the ambient water are very fast and, in general, are less than a few days. However, in saline conditions, we know little about the isotopic exchange between them.

The clay mineral assemblages in the Core SG-1 are shown in **Figure 3**. The curves of chlorite and illite abundances are similar, but they are different with those of the kaolinite, smectite, and I/Sm (**Figure 3**). The curve of smectite content shows a decreasing trend with decreasing depth, and that of the illite content shows an increasing trend slightly (**Figure 3**). The XRD analyses suggest that the I/Sm in the core exhibits order ($R_3$ I/Sm) (mixed layer of I/Sm > 80%). $R_1$ I/Sm is more stable than any other mixed layer I/Sm phase because it has a unique structure, composition, and order [17]. The clay minerals that were identified as $R_3$ I/Sm by XRD analyses are possible to be mixtures of discrete smectite and $R_1$ I/Sm [17].

The diagenesis of clay minerals in lacustrine sediments is a disputed topic. The clays in Core SG-1 in the Qaidam Basin probably underwent early diagenesis based on their curves and relationship between them (**Figures 4** and **5**) [18]. Climate changes, to some extent, control on the degree of alteration of the primary minerals and the composition of the secondary products. In the clay minerals, the exchange between the interlayer water and the ambient water recorded the diagenetic information. For instance, the $\boldsymbol{\delta}^{18}$O and $\boldsymbol{\delta}$D of the different clays and interlayer water record temperatures that range from surface temperatures to hydrothermal temperatures of about 150–200°C [19]. In situ precipitation of I/Sm minerals may take place in chemical and isotopic equilibrium with the reacting solution [20, 21]. However, in high saline conditions, the reaction or the above exchange may be extremely slowly.

### *2.2.2 The $\delta^{18}$O and $\delta$D of the interlayer water*

The $\delta^{18}$O of the interlayer water ranged from −11.8 to 82.95% with the average of 20.8%, and the δD ranged from −114.3 to 165.5% with the average of −35.6%

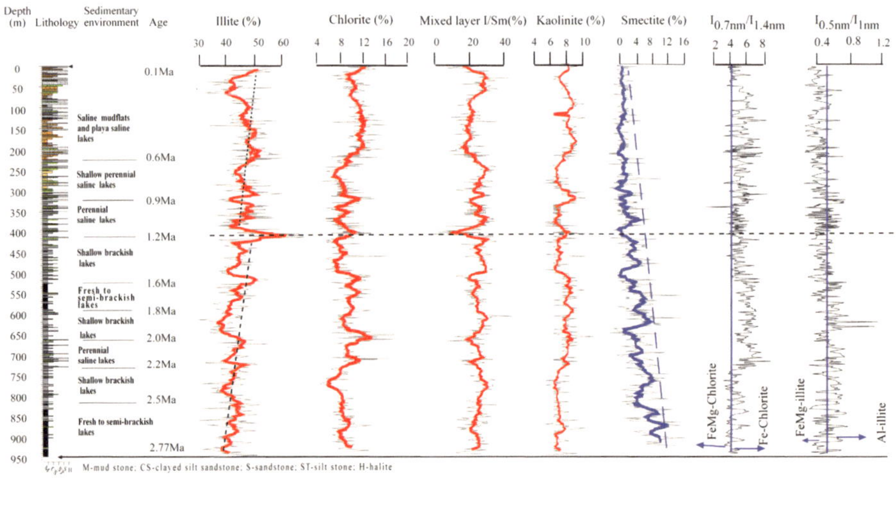


**Figure 3.**
*Variations in the amount of illite, chlorite, kaolinite, smectite, and illite/smectite (I/Sm) and in the chemistry of illite (I5Å/I10Å ratios) and chlorite (I7Å/I14Å ratios). Colored lines represent a five-point running average. Sedimentary environments are from Wang et al. [10]; ages are from Zhang et al. [9].*

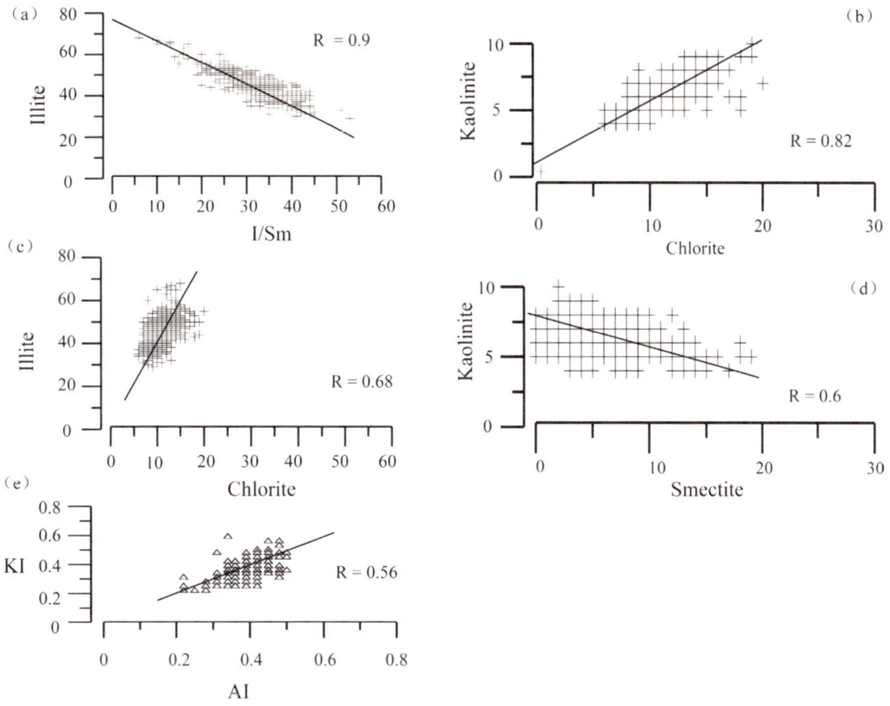


**Figure 4.**
*Correlations between the abundances (%) of (a) illite and I/Sm, (b) chlorite and kaolinite, (c) illite and chlorite, (d) smectite and kaolinite, (e) KI (Kübler index), and AI (Árkai index).*

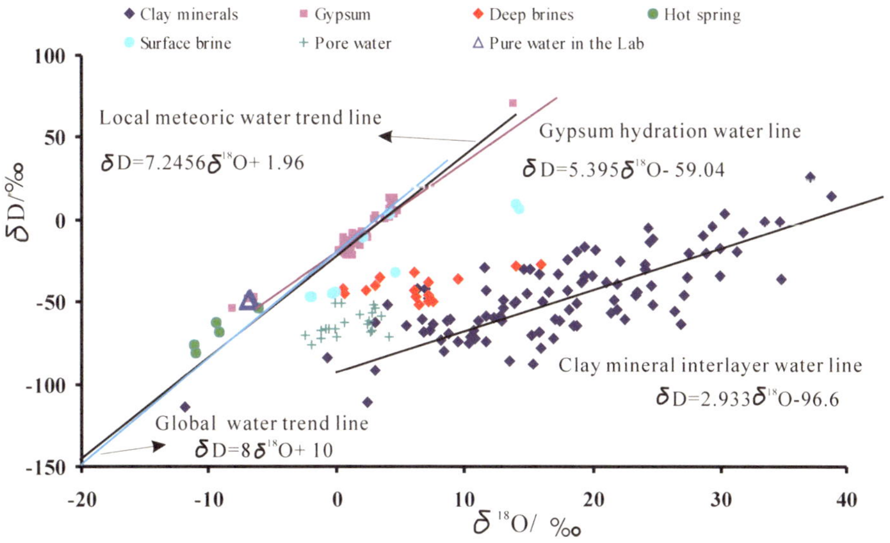


**Figure 5.**
*Relationship between the $\delta^{18}O$ and δD of the interlayer water of the clay minerals and different types of water from the Qaidam Basin. The global and modern local meteoric water lines are from Craig [22] and Li et al. [23], respectively. The present-day precipitation and hot spring data are from Zhang [24]. Gypsum hydration water data are from Li et al. [14], deep brine water data are from Fan et al. [25], and surface water and pore water data are from Xiao [26] and Fan et al. [25]. The large differences in the $\delta^{18}O$ and δD of the pure water and the interlayer water of the clay minerals suggest that the interlayer water did not exchange with the pure water and other water used during the processing and pretreatment.*

(**Figure 5**). They exhibit a roughly linear relationship and are significantly correlated ($R = 0.96$), indicating that these isotopes share the same sources. The isotopes of the interlayer water possibly recorded the brine composition which contained interstitial water or pore water derived from lake water and underwent strong evaporation. This could be understood as the following. (a) Some of the $\delta^{18}O$ and $\delta D$ of the interlayer water are close to those of surface water, deep brine, and pore water. (b) The rates of isotopic exchange between the interlayer water and the ambient water are very fast and, sometimes, are less than a few days. Therefore, the exchange rate between the structural oxygen of the clay minerals and that of the ambient water could be insignificant over a 2.8 Ma time span at a temperature of ≤100°C [27, 28]. (c) The slopes of the $\delta^{18}O$ and $\delta D$ of the interlayer water are smaller than that of the gypsum hydrated water that recorded the evolution of lake water or subsurface water (**Figure 5**), suggesting the interlayer water underwent more stronger evaporation than gypsum hydrated water. (d) The $\delta^{18}O$ and $\delta D$ of the interlayer water of clay minerals are very different with those of pure water (**Figure 5**), which was used to treat the clay minerals.

Independent of the oxygen isotopic exchange, the hydrogen isotopic exchange occurs by a proton exchange mechanism [28, 29]. The hydrogen isotopic exchange between smectite and its ambient water is very fast and significantly relative to that between illite and kaolinite with their ambient water [28].

## 2.3 Gypsum and the $\delta^{18}O$ and $\delta D$ of hydrated water

### *2.3.1 Gypsum*

Gypsum ($CaSO_4{\cdot}2H_2O$), one of the most abundant evaporite minerals occurring as a syndepositional evaporite, carries information about brines from which they precipitate. There are several shapes of gypsum crystals (**Figure 6**). These prismatic and tabular gypsum layers were vertically aligned, indicating primary subaqueous precipitates and with bottom nucleation [31, 32]. Some euhedral gypsum crystals were scattered in the mudstone-siltstone layers, which are possibly formed by evaporation of pore water or interstitial brine and regarded as diagenetic gypsum. Based solely on their morphology, it is not easy to confirm gypsum to be synsedimentary or diagenetic, because there is no diagnostic relationship between crystal morphology and depositional environment [33]. Besides the aligned gypsum, scattered euhedral crystals may be significant for understanding the evolution of lake water, because pore water or interstitial brine may have originated from infiltration of lake water. The Core SG-1 showed an increase of evaporite mineral contents upward, and the strata were mainly horizontal and not affected by tectonic deformation [10]. Therefore, these selected gypsums may have been primary in origin in arid, or aqueous, young lacustrine. These environments contain deposits that have never been exposed or deeply buried and may also retain their original isotopic compositions [34–36].

Minor cations such as $Sr^{2+}$, $K^+$, and $Mg^{2+}$ may be incorporated into gypsum lattice via substitution for $Ca^{2+}$ during the coprecipitation process because they have similar ionic radii. Different cations are selectively attracted to particular faces [37, 38]. For example, the (1 1 1) face is covered by either calcium ions or sulfate clusters and the (1 1 0) and (0 1 0) faces by both calcium and sulfate clusters [39]. $Na^+$ is preferentially adsorbed on (1 1 1) surface of sulfate salts that is dominantly occupied by $Ca^{2+}$ [39, 40]. However, the presence of $Na^+$ also impedes the (1 1 1) face growth and, as a result, minimizes parallel to its c axis [36, 39]. $K^+$ has a similar inhibitory effect [40]. Because of the difference in coordination number between $Ca^{2+}$ (eight in gypsum) and $Mg^{2+}$ (six in most minerals), $Mg^{2+}$ will disturb gypsum

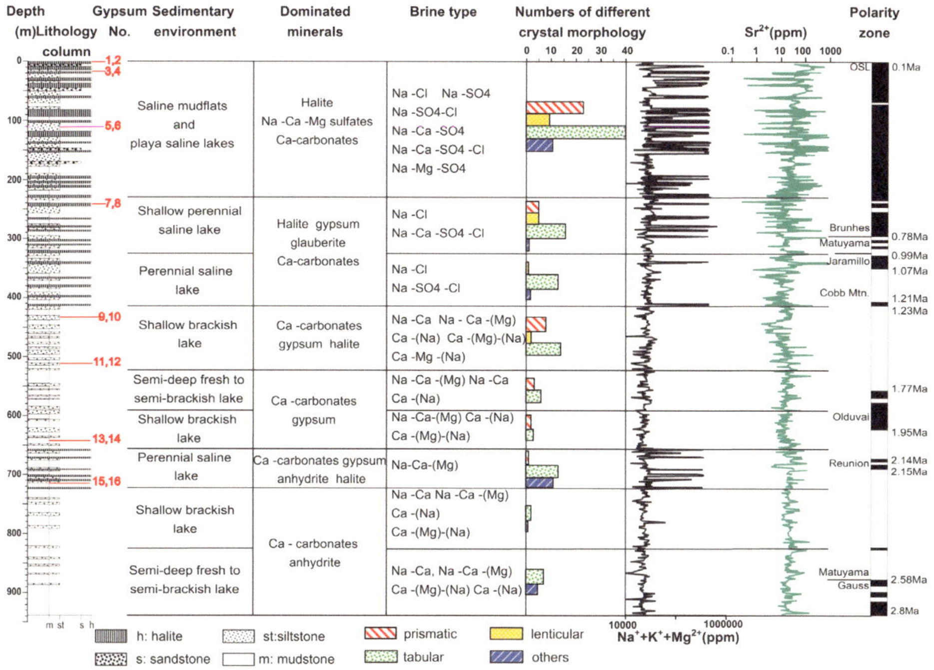


**Figure 6.**
*The monoclinic gypsum exhibits four principal forms in the Core SG-1 [30]: prismatic crystals, lenticular crystals, and tabular crystals. The tabular and prismatic crystals tend to increase with decreasing depth. The red numbers mean that the same kind of crystal morphology can appear in different environments and different crystal morphologies can also occur in the same depositional environment.*

growth along c axis [41]. Preferentially adsorbed on the (1 1 0) face, $Sr^{2+}$ forms epitaxial deposits of strontium sulfate [39]. Therefore, $K^{+}$, $Na^{+}$, and $Mg^{2+}$ make the c value smaller, while $Sr^{2+}$ reduces a and b values. As a result, there have different gypsum morphologies.

### *2.3.2 The $\delta^{18}O$ and $\delta D$ of hydrated water*

The $\delta^{18}O$ of gypsum hydrated water ranged from −4.21 to 8.69%, with average of 5.74%; and the δD was from −72.77 to 49.73%, with average of −28.09%. They exhibit a roughly linear relationship with the slope of 5.39, and its mother water-line has a slope of 5.52 (**Figure 7**). The $\delta^{18}O$ and δD appear to be much lower than those of today's mean precipitation ($\delta^{18}O$ = −9.25% and δD = −41.3%) [14]. This indicated a slightly weaker evaporation and/or colder climate than today [36, 43]. On the other hand, the $\delta^{18}O$ and δD are close to those of the Altyn Tagh Mountain groundwater that sourced from meteoric water (**Figure 7**). It is therefore likely that meteoric water was the main source of hydration water during gypsum formation.

In general, gypsum could be formed in the following ways: in situ formation (e.g., resulting from the oxidation of sulfide minerals); hydration of anhydrite; and direct deposition from an evaporating solution saturated with gypsum [44]. If gypsum was formed in situ formation, the $\delta^{18}O$ and δD of hydrated water would reflect those of meteoric and/or surface waters [44, 45]. If it is formed from the hydration of anhydrite, which is assumed to be a Rayleigh process, both the hydration water and its mother water would be expected to move along a line with a negative $\Delta\delta D/\Delta\delta^{18}O$ [46, 47]. In this study, the $\Delta\delta D/\Delta\delta^{18}O$ value was 5.39 [14], suggesting evaporation to be the major geochemical process for gypsum deposition, that is, the gypsum in the Core SG-1 was likely to deposit directly from brines. The degree

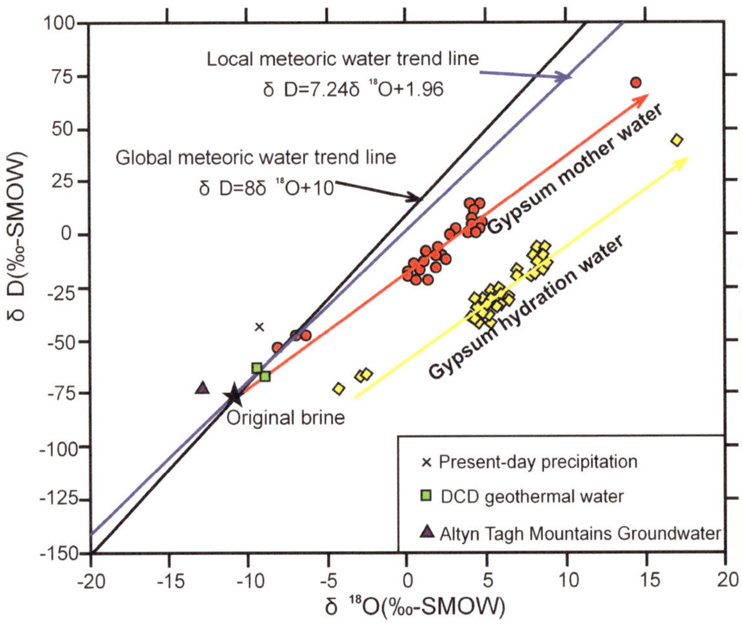


**Figure 7.**
*Relationship between $\delta^{18}O$ and $\delta D$ of gypsum hydration water and mother water as compared to meteoric water and different types of water from the Qaidam Basin. The global and modern local meteoric waterlines are from Craig [22] and Li et al. [23], respectively. The present-day precipitation and Dachaidan geothermal water are from Zhang [24]. Altyn Tagh Mountain groundwater data are from Wang et al. [42].*

of evaporation depends upon the salinity and atmospheric humidity [22, 48]. In humid conditions, $\Delta\delta D/\Delta\delta^{18}O$ gradients are usually between ~5 and ~6, whereas the gradient can be low as 2.5–4 in arid regions. The evaporation line gradient for present-day Qaidam Basin meteoric water is lower 4.4 [24], which is less than that of gypsum hydration water in Core SG-1. This suggested that brine in the 2.2–0.1 Ma had lighter isotopes than today's local meteoric water and that evaporation was weaker than today's.

## 2.4 Paleoclimatic implications

The climate in the area indicates a long-term persistent aridification event in inland Asia, which is supported by several climate proxies [10–15]. The $\delta^{18}O$ and $\delta D$ of the clay interlayer water and gypsum hydrated water both reflect the compositional variations of the lake water or pore water and record changes in environment [14, 18]. According to the significant positive relationship between $\delta^{18}O$ and $\delta D$ and their low slope (2.933, **Figure 7**), most of the interlayer water could be from pore water and lake water in an evaporative environment. The main factor contributing to the variations in the $\delta^{18}O$ and $\delta D$ water was probably E/P value oscillations, which were linked to the climate changes.

Both the lake water and gypsum hydrated water, their curves of the $\delta^{18}O$ and $\delta D$, display a stepwise increasing trend (**Figure 8**), indicating a drying trend from 2.2 to 0.1 Ma. This is consistent with the global cooling trend, which is also recorded by the $\delta^{18}O$ of marine sediments (**Figure 8**). The lower values of $\delta^{18}O$ and $\delta D$ in 1.2–0.1 Ma than in 2.77–1.2 Ma (**Figure 8**) agreed with the sedimentary changes from brackish lakes to saline and playa lakes [10]. Compared with brackish lakes, it is harder for brines in mudflat and playa lakes to lose water. The occurrence of $Na_2SO_4$-bearing salt minerals such as glauberite, thenardite, mirabilite, and bloedite [11] and the high $\delta^{18}O$ of carbonates [13] also suggest the climate to be extremely

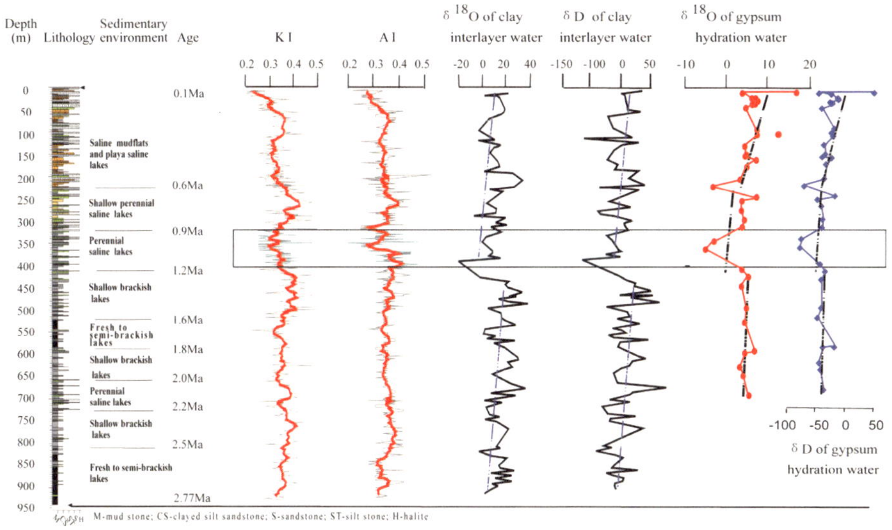


**Figure 8.**
*Hydrogen and oxygen isotopes of gypsum hydrated water and interlayer water of clay minerals vs. depth. The lithologic column and sedimentary environment are from Wang et al. [10]. The ages were based on paleomagnetic results of Zhang et al. [9].*

arid in 1.2–0.1 Ma. At about 1.2–1.1 Ma, the low value was likely due to the MPT cold event, which was also recorded in other places in the Qaidam Basin [49]. This is also consistent with other proxies in Central Asia such as lithological and sedimentary records, evaporative minerals, pollen records, grain size, and trace elements [10–13, 49–51].

In summary, $\delta^{18}O$ and δD of the gypsum hydrated water, and interlay water of clay in high saline conditions, both have recorded environmental signals.

## 3. The isotopes of Sr, S, and B of halite in saline conditions

Halite is common in evaporite deposits. Today, exploited evaporite deposits are most found in the arid and semiarid deserts of the world (the areas between latitudes 15° and 45° both north and south of the equator). Tectonics, climate, and origins of brine are the prime controls on most evaporite deposits. However, the origins of brine were a big topic. Giant evaporite deposits typically originate from brines with a marine or/and land origin(s), along with varying inputs from deeply circulated meteoric, basinal, and hydrothermal fluids [52]. The isotopes of S, Sr, and B of halite could be very useful in determining the origins of brine. We used well-known accepted methods to analyze halite and the isotopes to distinguish origins of brine for the evaporite deposits in Laos and Southern China.

### 3.1 Geological setting and the Core ZK2893

#### *3.1.1 Geological setting in Laos and Southern China*

Southeast Asia is composed of a series of Gondwana-derived continental blocks which experienced heterogeneous collision with the closure of multiple Tethyan Ocean branches (see reviews in Metcalfe [53, 54]). The Khorat Basin (KB) in Thailand and Sakhon Nakhon Basin (SNB) and the Lanping-Simao Basin (LSB) in Southwest China belong to the Indochina Block (**Figure 9**). The central part of the

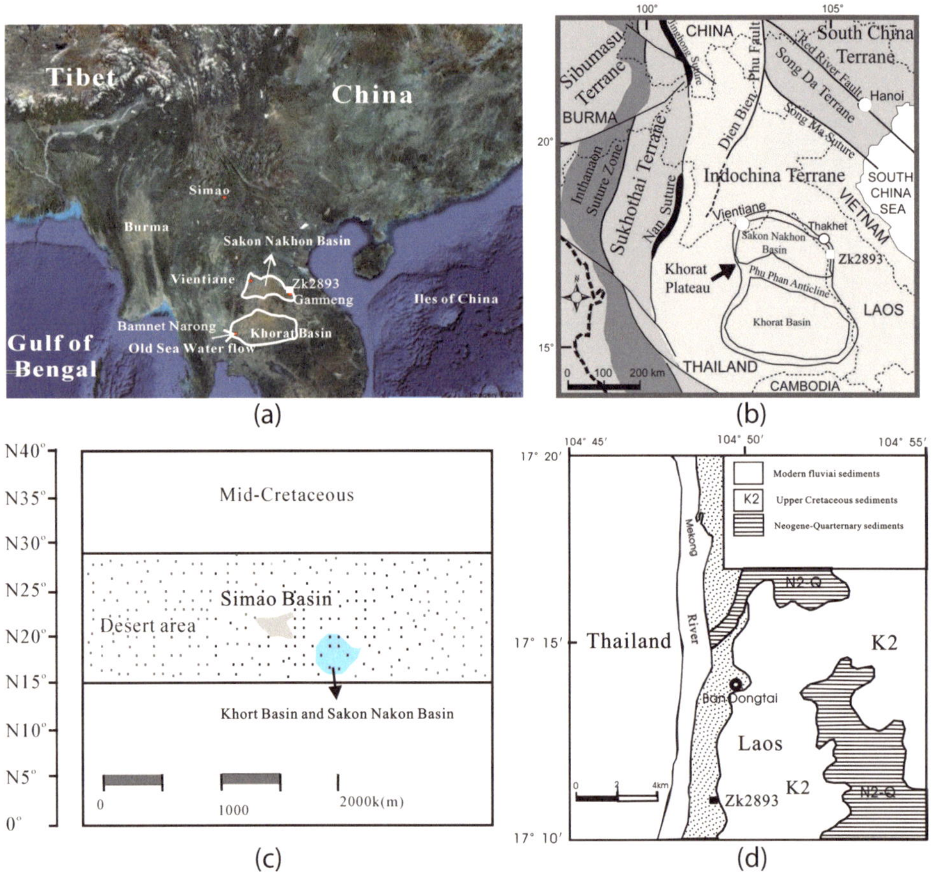


**Figure 9.**
*(a) Google Earth map showing the locations of the Sakhon Nakhon Basin and the Khorat Basin; (b) tectonic map modified from Sone and Metcalfe [55] and El Tabakh et al. [56]; (c) location of Khorat Basin and Sakhon Nakhon Basin, in the Middle and Late Cretaceous (after El Tabakh et al. [56]); (d) geological map showing the general distribution of sediments in the study region (after Zhang et al. [57]).*

Indochina Block (the Khorat Plateau) is relatively rigid with respect to the Simao Terrane (including the LSB) and has been experiencing clockwise rotation of about 15° up to the present day [58, 59]. The location of LSB changed from 25.7° N in Cretaceous to 18.6°N in Paleocene and Eocene [60]. This suggested that the LSB was gradually southward approaching the KB and SNB.

During the Middle and Late Cretaceous, salts formed; however, these salt formations lie atop thick nonmarine sediments of the Mesozoic Khorat Group. The nonmarine sediments were more than 5 km thick, which were deposited in the Late Triassic [56, 61, 62]. From the very Late Triassic to the Early Cretaceous, the KB and SNB were filled with fluvial and lacustrine facies [56]. Due to the occurrence of similar salt minerals within the same tectonic belt in KB and the LSB, they have formed in brines from similar sources [63].

In the mid-Cretaceous, the basins (LSB, KB, and SNB) were within the subtropical high-pressure belt [64]. However, the scale of the Mengye potash deposit in the LSB is much smaller than those in the KB and SNB of Thailand and Laos [63]. This surprising contrast drove many scientists to explore it in the area for ~50 years. But no consensus has therefore been reached regarding their fluid origins [56, 65, 66].

### *3.1.2 The Core ZK2893 in Laos*

The lithostratigraphy contained three evaporite-clastic cycles, upper member (148.4 m to top), middle member (299.2–148.4 m), and lower member cycles (595.4–299.2 m), which are composed of evaporite unit and red-colored siliciclastic unit (**Figure 10**). The lithostratigraphy of the core could match with a reviewed section by El Tabakh et al. [56]. Blocky halite beds were only present in the lower and middle members (**Figure 11**). Chevron or cumulate crystals were well developed. Some thin salt units are observed (**Figure 10**), suggesting salt dissolution to be happened and their dissolved solutes to be mixed with their adjacent mudstone. Salt dissolution resulted in a lack of salts in many cores and the deposition of anhydrite [56].

### 3.2 Halite

The evaporite minerals in the halite beds are pure, massive red halite in ZK2893 (**Figure 11**). Some samples have trace or minor anhydrite content [60]. Compared with gypsum, the crystal shapes of halite seldom vary. But halite is much easier to be dissolved and recrystallized than gypsum. The halite crystals in the Core ZK2893 were primary, because (a) the primary fluid inclusions were developed (**Figure 12**), suggesting the halite was not dissolved; (b) the curve of Br in the core changed with depth (**Figure 13C**), consistent with the Br curve of primary halite (**Figure 13A**) and very different with the Br curve of secondary halite (**Figure 13B**); and (c) the Br of basal halite was the lowest in the core, suggesting the primary halite to have been preserved.

Basal halite is defined as the first Cl mineral to be precipitated during the evaporation of seawater/lake water. During syndepositional and early diagenetic processes, Br content of basal halite is stable [69, 71], and the initial compositions of

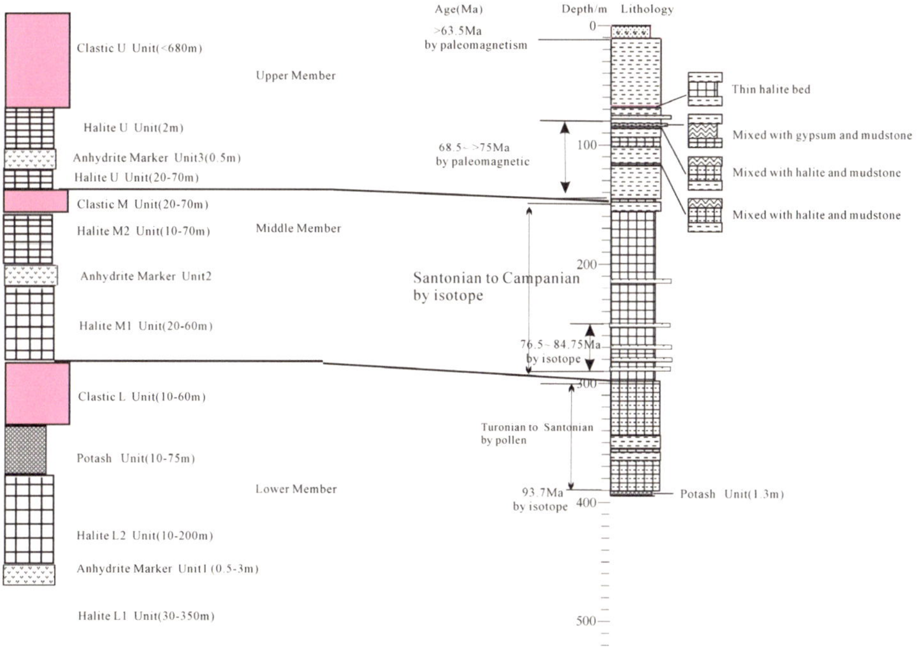


**Figure 10.**
*Left: lithostratigraphical review of the evaporite formation of the Khorat basin and Sakhon Nakhon Basins, after El Tabakh et al. [56]. Right: lithostratigraphy of Core ZK2893 (this study). The paleomagnetic ages are from Zhang [67]; the pollen ages are from Zhong et al. [62]; the isotopic ages are from Hansen et al. [68].*

**Figure 11.**
*Photos of halite in Core ZK2893 (after Li et al. [60]).*

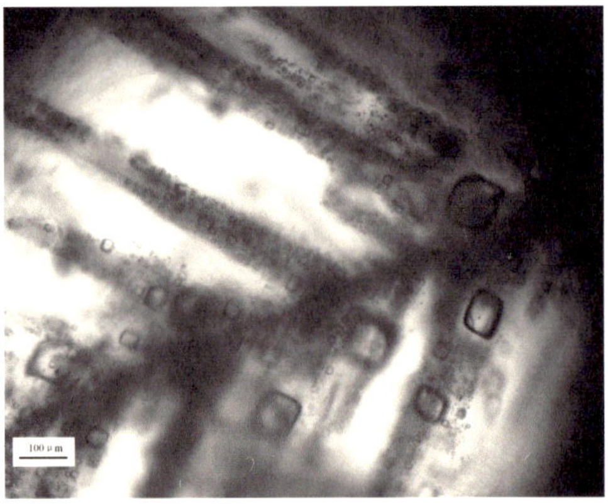


**Figure 12.**
*Photomicrographs showing primary fluid inclusion banding in chevron halite.*

brine are well-preserved in basal halite [60, 71]. Therefore, the Br content of basal halite can be used to reconstruct the composition and origins of paleo-water. For example, Siemann [71] used the Br content of primary basal halite to reconstruct the Br variations in seawater over the past 500 Ma.

## 3.3 The isotopes of sulfur, strontium, and boron

Isotopic compositions of sulfur, boron, and strontium could be more robust indicators of the origins of evaporites (marine vs. continental origins).

### *3.3.1 Sulfur isotope*

Waters from continental sources in general have lower $\delta^{34}S$ values than those from contemporary seawater. The $^{34}S$ values of freshwater usually range from −5 to 5%, while that of recent seawater is globally uniform with value being 20 ± 0.5% [72]. Sulfate minerals (such as gypsum and anhydrite) have stable $\delta^{34}S$ values and are typically resistant to diagenetic alteration [72–74]. Factors influencing the $\delta^{34}S$ values of evaporative sulfates are marine and nonmarine contributions, reservoirs, and redox [72, 74]. For example, bacterial-facilitated sulfate reduction is preferentially enriched in lighter $^{32}S$ isotope, and therefore, the residual sulfate minerals

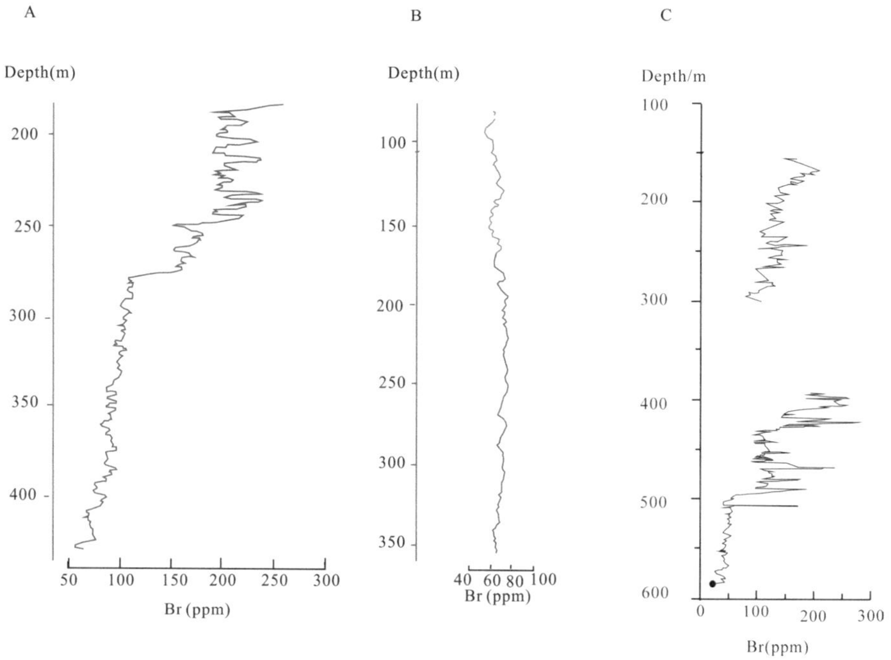


**Figure 13.**
*(A) Curves of Br in primary halite in KB, after Hite and Japakasetr [69]. (B) Curves of Br in secondary halite, after Hite and Japakasetr [69]. (C) Curves of Br in the Core ZK2893 in SNB, Laos [70], suggesting the halite in the core to be primary.*

(such as gypsum) will enrich heavy $^{34}S$ [72, 73]. In contrast, reservoir effects can decrease $\delta^{34}S$ values of the subsequently precipitated gypsum. If sulfate evaporites precipitated in a closed system, the $\delta^{34}S$ values expectedly display a gradual decrease upward in a vertical stratigraphic sequence. The reservoir effect played an active role influencing the $\delta^{34}S$ values of gypsum, anhydrite, and halite in the abovementioned basins (**Figure 14**).

The average of $\delta^{34}S$ values increase from the lower to the upper member, which indicates an evaporation effect. Microbial effect on the core's $\delta^{34}S_{-SO4}$ values may be insignificant. This could be understood in the following. (a) There is no relationship between sulfate concentrations and the $\delta^{34}S_{-SO4}$ values of halite (**Figure 14**). Generally, microbial sulfate reduction can result in an inverse correlation of them [78]. And (b) the $\delta^{34}S$ values observed in the halite and anhydrite are very similar. Redox recycling of bacterial reduction can result in the enrichment of $\delta^{34}S$ values and significant difference within evaporites [72, 73]. Therefore, $\delta^{34}S$ in this study could be used to identify the origins of brine. In the Core ZK2893, all of $\delta^{34}S$ values are lower than, or similar to, those of contemporary seawater (**Figure 14**), suggesting a possible continental origin. Both the $\delta^{34}S$ and $\delta^{18}O_{-SO4}$ values of halite are lower than those of Middle Cretaceous seawater (**Figure 14**) with a short range dispersion, shorter than that between $\delta^{34}S$ and $SO_4$ values (**Figure 14**), suggesting that paleo-lake water possible to be mixed with freshwater.

### *3.3.2 Strontium isotope*

Strontium is a divalent alkaline earth element. Its isotope composition is almost homogeneous with modern seawater with $^{87}Sr/^{86}Sr$ values of 0.709175–0.709235 [79, 80]. Because the Sr composition/isotope ratios for the weathering and erosion of rocks are high [81], the $^{87}Sr/^{86}Sr$ ratios for river water evince a large range

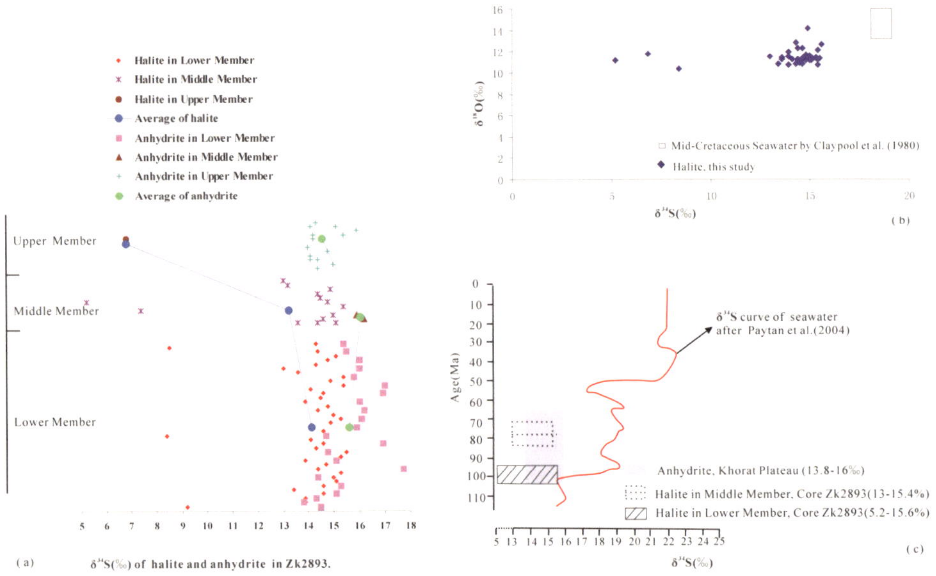


**Figure 14.**
*(a) $\delta^{34}S$ values of halite for Core ZK2893 and anhydrite. Anhydrite data are from El Tabakh et al. [58] and Xu [75]. (b) Plot of $\delta^{34}S$-$\delta^{18}O_{-SO_4}$ of halite and Mid-Cretaceous seawater. Mid-Cretaceous seawater data are from Claypool et al. [76]. (c) Comparison of $\delta^{34}S$ values between halite in this study and seawater. The $\delta^{34}S$ curve of seawater from Paytan et al. [77].*

from 0.7074 to 0.803 (**Figure 15**). In a continental setting, the $^{87}Sr/^{86}Sr$ ratio for continental waters is almost always higher than that for marine waters (**Figure 15**). The $^{87}Sr/^{86}Sr$ values of Paleocene halite in the Qaidam Basin (a continental basin), for example, were higher than those of Paleocene seawater (**Figure 9**). Thus, the $^{87}Sr/^{86}Sr$ values of authigenic minerals are very variable and higher than those of seawater when the minerals are precipitated from a solution.

As no fractionation of Sr isotopes occurs during halite precipitation, the measured halite $^{87}Sr/^{86}Sr$ ratios for Core ZK2893 (ranging from 0.707443 to 0.708587) could represent those of the parent solution. $^{87}Sr/^{86}Sr$ ratios for the lower member ranged from 0.707443 to 0.708587, averaging 0.707693, while those for the middle member ranged from 0.70752 to 0.708105, averaging 0.70764. There was only one $^{87}Sr/^{86}Sr$ ratio value in the upper member, of 0.708163. In the lower member, the possible reasons for the high $^{87}Sr/^{86}Sr$ ratios in the upper part (**Figure 15**) are (a) the permeation of continental waters from the middle clastic unit into the halite layers and (b) an inflow of continental waters into the Sakhon Nakhon Basin during the shrinkage process of the basin. However, the $^{87}Sr/^{86}Sr$ ratios exhibit no significant variability in the middle member [83], suggesting no, or little, penetration of continental waters into the halite layers. The $^{87}Sr/^{86}Sr$ and 1/Sr halite ratios in Core ZK2893 do not show a linear relation but rather a weak positive relation (**Figure 15a**). Some $^{87}Sr/^{86}Sr$ ratios are far higher than those of contemporary seawater, indicating the possible input of continental waters into these evaporite basins; these $^{87}Sr/^{86}Sr$ ratios in the lower member are lower than, or similar to, those of Cretaceous seawater; further, mean Middle Cretaceous $^{87}Sr/^{86}Sr$ ratios are similar to those of paleo-seawater (**Figure 15b**), suggesting a marine remnant origin.

### *3.3.3 Boron isotope*

Natural boron, a highly soluble element, has two stable isotopes ($^{10}B$ and $^{11}B$) with approximate abundances of 20 and 80%, respectively. Boron isotope

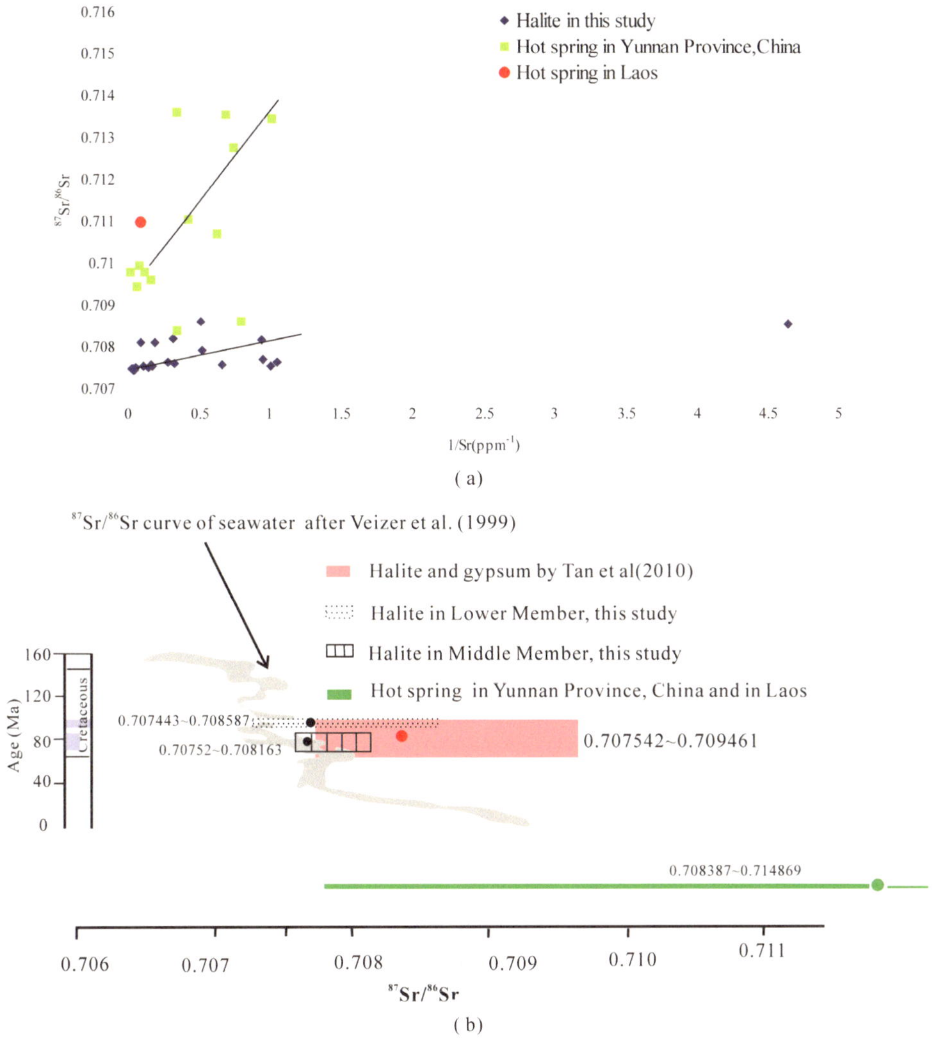


**Figure 15.**
*(a) Plot of $^{87}Sr/^{86}Sr$ and 1/Sr of halite and hot spring water; and (b) $^{87}Sr/^{86}Sr$ ratios in halite, gypsum, hot spring water and saline lake water: A dot indicates a mean value. Gypsum data are from Tan et al. [66], hot spring data for Yunnan Province in China, and for Laos, are from Bo et al. [82].*

fractionation is dependent upon the distribution of boron species, temperature, and pH, but pH is the key factor [84–89]. Dissolved boron exists mainly in the form of $B(OH)_3$ and $B(OH)_4$, which are dominantly present as $^{11}B(OH)_3$ at low pH values and as $^{10}B(OH)_4$ at high pH values [85–88]. Because pH values increased with the increasing extent of evaporation, the minerals precipitated in early stages have higher $\delta^{11}B$ values than those of minerals precipitated in later stages. Generally, the sequence of $\delta^{11}B$ values of different minerals is following as: carbonate > gypsum > borate > halite > sylvite. As boron was present in inclusion [90], the $\delta^{11}B$ values of halite and sylvite are exactly representative of those for contemporary paleo-brines. Therefore, the $\delta^{11}B$ values of halite and sylvite are higher than those in carbonates and borate minerals, and all of minerals have lower $\delta^{11}B$ than those of brine. In the Core ZK2893, however, the $\delta^{11}B$ values of halite and sylvite were lower than those of borate minerals (**Figure 16**). One possible reason is that there have different sources of brines, such as continental, marine residue and groundwater/hydrothermal brine. In the LSB, China, most of $\delta^{11}B$ values of

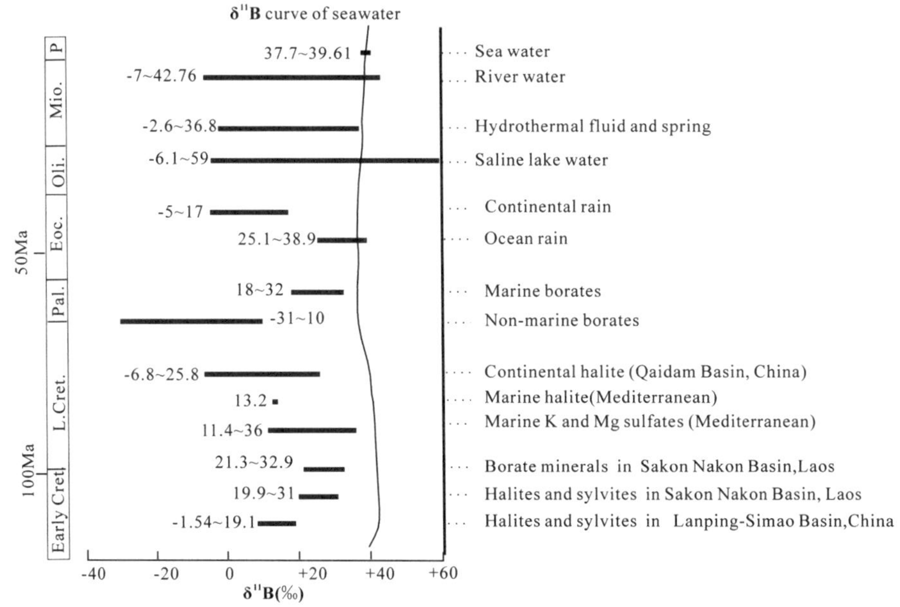


**Figure 16.**
*$\delta^{11}$B values from published papers. Marine and nonmarine borates after Swihart et al. [91], Xiao et al. [90], and Palmer and Helvaci [85]. Seawater after Vengosh et al. [86] and Foster et al. [89]. River water after Vengosh et al. [7]; Rose et al. [92]; and Lemarchand et al. [93]. Halite after from Vengosh et al. [86]; Liu et al. [94]; and Kloppmann et al. [95]. Borate minerals in SNB after Zhang et al. [96]. Halite and sylvite in SNB after Tan et al. [66]). Halite and sylvite in LSB after Zhang et al. [97]. Curve of seawater after Lemarchand et al. [98].*

halite and sylvite ranged from −1.54 to 19.1%, which are lower than those in Laos (19.9–31.1‰) (**Figure 16**). This suggested the paleo-brines in LSB, China, are from Laos and diluted during the flow process.

Comparison of $\delta^{11}$B values between known marine minerals and unknown minerals is a useful way to distinguish marine and continental origin. In nonmarine setting, some $\delta^{11}$B values of hydrothermal fluid, saline lake water and river water, are higher than those of seawater (**Figure 16**), that is, lower $\delta^{11}$B values of minerals than that of seawater are not indicate continental origin. The comparison of $\delta^{11}$B between salt minerals is valid. Although the $\delta^{11}$B values of halite in Laos were lower than those of seawater, they were near to those of marine borates (**Figure 16**). This suggested minor or trace marine origin. The $\delta^{11}$B values of the evaporite in LSB in China are lower than, but some are near to, those of halite and sylvite in SNB, Laos, also suggesting continental origin with minor residual seawater.

In summary, halite isotopic compositions (in the form of $\delta^{34}$S-$\delta^{18}O_{-SO4}$ values, $^{87}$Sr/$^{86}$Sr and $\delta^{11}$B) appear to represent an original composition associated with brine. These isotopic proxies indicate that continental and hydrothermal origins are likely to be more important sources of evaporite deposits than marine origins for these Laotian evaporite deposits.

## 4. Summary

The gypsum in the Core SG-1 was deposited directly from brines and was stable after deposition. The $\delta^{18}$O and δD curves of hydrated water in gypsum record the evolution of paelo-lake in Qaidam Basin, Western Tibetan Plateau since 2.2 Ma, including the famous cold event, mid-Pleistocene event.

In high saline conditions, the clay minerals from Core SG-1 may have undergone early diagenesis. The primary illite and chlorite contents in Core SG-1 were lower than those observed, while the primary smectite and kaolinite contents were higher than those observed. After early diagenesis, there were no isotopic exchange between the interlayer water and the ambient water. The $\delta^{18}O$ and $\delta D$ values indicated the interlayer water of clay minerals to be more concentrated than those of hydrated water. The isotopic composition of the interlayer water reflects variations in the pore water/lake water or in the reacting solutions and also records changes in environment.

Using isotopic compositions of salt minerals to identify origins of brine is a complex topic. It should be careful to use sole isotope to identify the origins. For the evaporite deposits in LSB in China and NSB, in Laos, these isotopic proxies, including $\delta^{34}S$-$\delta^{18}O_{-SO4}$, $^{87}Sr/^{86}Sr$, and $\delta^{11}B$, indicate that continental and hydrothermal origins are likely to be more important than marine origins.

## Acknowledgements

This study was supported by the Strategic Priority Research Program of Chinese Academy of Sciences (Grant No. XDA20070201, XDA20070101), the National Basic Research Program of China (Grant No. 2017YFC0602803), the International Cooperation Project of the Chinese Academy of Sciences (Grant No. 131C11KYSB20160072), the Second Tibetan Plateau Scientific Expedition and Research (Grant No. 2019QZKK0202), and the National Natural Science Foundation of China (Grant No. 41620104002).

## Author details

Minghui Li[1,2]*, Xiaomin Fang[2,3,4], Jiao Li[1], Maodu Yan[2,3], Shurui Sun[1] and Liping Zhu[1,2,4]

1 Key Laboratory of Tibetan Environment Changes and Land Surface Processes, Institute of Tibetan Plateau Research, Chinese Academy of Sciences (ITPCAS), Beijing, China

2 CAS Center for Excellence in Tibetan Plateau Earth Sciences, Beijing, China

3 Key Laboratory of Continental Collision and Plateau Uplift, ITPCAS, Beijing, China

4 University of Chinese Academy of Sciences, Beijing, China

*Address all correspondence to: liminghui@itpcas.ac.cn

## References

[1] Chen K, Bowler JM. Late Pleistocene evolution of salt lakes in the Qaidam basin, Qinghai province, China. Palaeogeography, Palaeoclimatology, Palaeoecology. 1986;**54**:87-104

[2] Qian ZQ, Qu YH, Liu Q, editors. Potash Deposits. Beijing: Geological Press; 1994 (in Chinese)

[3] Fang X, Zhang W, Meng X, Gao J, Wang X, King J, et al. High-resolution magnetostratigraphy of the Neogene Huaitoutala section in the eastern Qaidam Basin on the NE Tibetan Plateau, Qinghai Province, China and its implication on tectonic uplift of the NE Tibetan Plateau. Earth and Planetary Science Letters. 2007;**258**(1-2):293-306

[4] Tuo J, Philp RP. Occurrence and distribution of high molecular weight hydrocarbons in selected non-marine source rocks from the Liaohe, Qaidam and Tarim Basins, China. Organic Geochemistry. 2003;**34**:1543-1558

[5] Liu X, Dong H, Jason AR, Matsumoto R, Yang B, Wang Y. Evolution of Chaka Salt Lake in NW China in response to climatic change during the Latest Pleistocene-Holocene. Quaternary Science Reviews. 2008;**27**:867-879

[6] Mischke S, Sun Z, Herzschuh U, Qiao Z, Sun N. An ostracod-inferred large Middle Pleistocene freshwater lake in the presently hyper-arid Qaidam Basin (NW China). Quaternary International. 2010;**218**:74-85

[7] Vengosh A, Chivas AR, Starinsky A, Kolodny Y, Baozhen Z, Zhang P. Chemical and boron isotope compositions of non-marine brines from the Qaidam Basin, Qinghai, China. Chemical Geology. 1995;**120**:135-154

[8] Eugster HP. Geochemistry of evaporitic lacustrine deposits. Annual Review of Earth and Planetary Sciences. 1980;**8**:35-63

[9] Zhang WL, Appel E, Fang XM, Song CH, Cirpka O. Magnetostratigraphy of deep drilling core SG-1 in the western Qaidam Basin (NE Tibetan Plateau) and its tectonic implications. Quaternary Research. 2012;**78**(1):139-148

[10] Wang JY, Fang XM, Appel E, Song CH. Pliocene-Pleistocene climate change at the NE Tibetan Plateau deduced from lithofacies variation in the drill Core SG-1, western Qaidam basin, China. Journal of Sedimentary Research. 2012;**82**(12):933-952

[11] Li MH, Fang XM, Wang JY, Song YG, Yang YB, Zhang WL, et al. Evaporite minerals of the lower 538.5 m sediments in a long core from the Western Qaidam basin, Tibet. Quaternary International. 2013;**298**:123-133

[12] Li MH, Fang XM, Yi CL, Gao SP, Zhang WL, Galy A. Evaporite minerals and geochemistry of the upper 400 m sediments in a core from the Western Qaidam Basin, Tibet. Quaternary International. 2010;**218**(1-2):176-189

[13] Han WX, Fang XM, Ye CC, Teng XH, Zhang T. Tibet forcing quaternary stepwise enhancement of westerly jet and central Asian aridification: Carbonate isotope records from deep drilling in the Qaidam salt playa, NE Tibet. Global and Planetary Change. 2014;**116**:68-75

[14] Li J, Li MH, Fang XM, Zhang GX, Zhang WL, Liu XM. Isotopic composition of gypsum hydration water in deep Core SG-1, western Qaidam basin (NE Tibetan Plateau), implications for paleoclimatic evolution. Global and Planetary Change. 2017;**155**:70-77

[15] Yang YB, Fang XM, Galy A, Li MH, Appel E, Liu XM. Paleoclimatic significance of rare earth element record of the calcareous lacustrine sediments from a long core (SG-1) in the western Qaidam Basin, NE Tibetan plateau. Journal of Geochemical Exploration. 2014;**145**:223-232

[16] Singer A. The paleoclimatic interpretation of clay minerals in sediments: A review. Earth-Science Reviews. 1984;**21**:251-293

[17] Dong H. Interstratified illite-smectite: A review of contributions of TEM data to crystal chemical relations and reaction mechanisms. Clay Science. 2005;**12**(Suppl. 1):6-12

[18] Li M, Sun S, Fang X, Wang C, Wang Z, Wang H. Clay minerals and isotopes of Pleistocene lacustrine sediments from the western Qaidam Basin, NE Tibetan Plateau. Applied Clay Science. 2018;**162**:382-390

[19] Savin SM, Lee M. Isotopic studies of phyllosilicates. In: Bailey SW, editor. Hydrous Phyllosilicates (Exclusive of Micas). Reviews in Mineralogy. Vol. 19. Washington, DC: Mineralogical Society of America; 1988. pp. 189-224

[20] Inoue A, Meunier A, Beaufort D. Illite-smectite mixed-layer minerals in felsic volcanic lastic rocks from drill cores, Kakkonda, Japan. Clays and Clay Minerals. 2004;**52**(1):66-84

[21] Sánchez-Roa C, Jiménez-Millán J, Abad I, Faulkner DR, Nieto F, García-Tortosa FJ. Fibrous clay mineral authigenesis induced by fluid-rock interaction in the Galera fault zone (Betic Cordillera, SE Spain) and its influence on fault gouge frictional properties. Applied Clay Science. 2016;**134**:275-288

[22] Craig H. Isotopic variations in meteoric waters. Science. 1961;**133**(3465):1702-1703

[23] Li X, Zhang M, Li Y, Wang S, Huang X, Ma Q, et al. Characteristics of δ18O in precipitation and moisture transports over the arid region in Northwest China. Environmental Sciences. 2012;**33**:711-719 (in Chinese with English abstract)

[24] Zhang P. Salt Lake of Qaidam Bain. Beijing: Science Press; 1987. pp. 32-233 (in Chinese)

[25] Fan Q, Ma HZ, Tan HB, Xu JX, Li TW. Characteristics and origins of brines in Western Qaidam Basin. Geochimica. 2007;**36**(6):633-637

[26] Xiao Y. Oxygen and hydrogen isotope research of different waters in Qarhan salt lake and lake sediments. Journal of Xiamen University (Natural Science). 1995;**34**(2):255

[27] James AT, Baker DR. Oxygen isotope exchange between illite and water at 22°C. Geochimica et Cosmochimica Acta. 1976;**40**:235-239

[28] O'Neil JR, Kharaka YK. Hydrogen and oxygen isotope exchange reactions between clay minerals and water. Geochimica et Cosmochimica Acta. 1976;**40**:241-246

[29] Zheng YF, Chen JF. Stable Isotope Geochemistry. Beijing: Science Publishing House; 2000. pp. 218-247 (in Chinese)

[30] Li J, Li MH, Fang XM, Wang ZR, Zhang WL, Yang YB. Variations and mechanisms of gypsum morphology along deep core SG-1, western Qaidam Basin (northeastern Tibetan Plateau). Quaternary International. 2017;**430**:71-81

[31] Warren JK. Evaporites brines and base metals: What is an evaporite? Defining the rock matrix. Australian Journal of Earth Sciences. 1996;**43**:115-132

[32] Gibert L, Ortí F, Rosell L. Plio-pleistocene lacustrine evaporites of the Baza Basin (Betic Chain, SE Spain). Sedimentary Geology. 2007;**200**:89-116

[33] Mees F, Casteneda C, Herrero J, Ranst EV. The nature and significance of variations in gypsum crystal morphology in dry lake basins. Journal of Sedimentary Research. 2012;**82**:37-52

[34] Hodell DA, Turchyn AV, Wiseman CJ, Escobar J, Curtis JH, Brenner M, et al. Late glacial temperature and precipitation changes in the lowland Neotropics by tandem measurement of δ18O in biogenic carbonate and gypsum hydration water. Geochimica et Cosmochimica Acta. 2012;**77**:352-368

[35] Evans NP, Turchyna AV, Gázquez F, Bontognali TRR, Chapman HJ, Hodell DA. Coupled measurements of δ18O and δD of hydration water and salinity of fluid inclusions in gypsum from the Messinian Yesares Member, Sorbas Basin (SE Spain). Earth and Planetary Science Letters. 2015;**430**:499-510

[36] Khademi H, Mermut AR, Krouse HR. Isotopic composition of gypsum hydration water in selected landforms from central Iran. Chemical Geology. 1997;**138**(3):245-255

[37] Cody RD. Organo-crystalline interactions in evaporite systems: The effects of crystallization inhibition. Journal of Sedimentary Research. 1991;**61**:704-718

[38] Magee JW. Late quaternary lacustrine, groundwater, aeolian and pedogenic gypsum in the Prungle Lakes, southeastern Australia. Palaeogeography, Palaeoclimatology, Palaeoecology. 1991;**84**:3-42

[39] Edinger SE. An investigation of the factors which affect the size and growth rates of the habit faces of gypsum. Journal of Crystal Growth. 1973;**18**:217-224

[40] Zhang J, Nancollas GH. Influence of calcium/sulfate molar ratio on the growth rate of calcium sulfate dihydrate at constant supersaturation. Journal of Crystal Growth. 1992;**118**:287-294

[41] Franchini-Angela M, Rinaudo C. Influence of sodium and magnesium on the growth-morphology of gypsum, $CaSO_4$. $2H_2O$. Neues Jahrbuch Fur Mineralogie Abhandlungen. 1989;**160**:105-115

[42] Wang ML, Yang ZC, Liu CL, Xie ZC, Jiao PC, Li CH. Potash Deposits and Their Exploitation Prospects of Saline Lakes of the Northern Qaidam Basin. Beijing: Geological Publishing House; 1996. pp. 33-51 (in Chinese)

[43] Farpoor MH, Khademi H, Eghbal MK, Krouse HR. Mode of gypsum deposition in southeastern Iranian soils as revealed by isotopic composition of crystallization water. Geoderma. 2004;**121**(3-4):233-242

[44] Sofer Z. Isotopic composition of hydration water in gypsum. Geochimica et Cosmochimica Acta. 1978;**42**(8):1141-1149

[45] Dowuona GN, Mermut AR, Krouse HR. Isotopic composition of hydration water in gypsum and hydroxyl in jarosite. Soil Science Society of America Journal. 1992;**56**(1):309-313

[46] Matsubaya O, Sakai H. Oxygen and hydrogen isotopic study on the water of crystallization of gypsum from the Kuroko type mineralization. Geochemical Journal. 1973;**7**(3):153-165

[47] Chen F, Turchyn AV, Kampman N, Hodell D, Gázquez F, Maskell A, et al. Isotopic analysis of sulfur cycling and gypsum vein formation in a natural $CO_2$ reservoir. Chemical Geology. 2016;**436**:72-83

[48] Sofer Z, Gat JR. The isotope composition of evaporating brines: Effect of the isotopic activity ratio in saline solutions. Earth and Planetary Science Letters. 1975;**26**(2):179-186

[49] Cai MT, Fang XM, Wu FL, Miao YF, Appel E. Pliocene-Pleistocene stepwise drying of Central Asia: Evidence from paleomagnetism and sporopollen record of the deep borehole SG-3 in the western Qaidam Basin, NE Tibetan Plateau. Global and Planetary Change. 2012;**94-95**:72-81

[50] Wu FL, Fang XM, Ma YZ, Herrmann M, Mosbrugger V, An ZS, et al. Plio–Quaternary stepwise drying of Asia: Evidence from a 3-Ma pollen record from the Chinese Loess Plateau. Earth Planetary Science Letters. 2007;**257**(1-2):160-169

[51] Ding ZL, Sun JM, Liu DS. Stepwise advance of the Mu Us desert since late Pliocene: Evidence from a red clay-loess record. Chinese Science Bulletin. 1999;**44**:1211-1214

[52] Warren JK. Evaporites: Sediments, Resources and Hydrocarbons. Berlin: Springer; 2006

[53] Metcalfe I. Tectonic framework and Phanerozoic evolution of Sundaland. Gondwana Research. 2011;**19**:3-21

[54] Metcalfe I. Gondwana dispersion and Asian accretion: Tectonic and palaeogeographic evolution of eastern Tethys. Journal of Asian Earth Sciences. 2013;**66**:1-33

[55] Sone M, Metcalfe I. Parallel Tethyan sutures in mainland Southeast Asia: New insights for Palaeo-Tethys closure and implications for the Indosinian orogeny. Comptes Rendus Geoscience. 2008;**340**(2-3):166-179

[56] El Tabakh M, Utha-Aroon C, Schreiber BC. Sedimentology of the Cretaceous Maha Sarakham evaporites in the Khorat Plateau of northeastern Thailand. Sedimentary Geology. 1999;**123**(1-2):31-62

[57] Zhang XY, Ma HZ, Tan HB, Gao DL, Li BK, Wang MX, et al. Preliminary studies of on geochemistry and post-depositional change of Dong Tai potash deposit in Laos. Mineral Deposits. 2010;**4**:713-721

[58] Yang ZY, Besse J. Paleomagnetic study of Permian and Mesozoic sedimentary rocks from Northern Thailand supports the extrusion model for Indochina. Earth and Planetary Science Letters. 1993;**117**:525-552

[59] Yang W, Spencer RJ, Roy Krouse H, Lowenstein TK, Casas E. Stable isotopes of lake and fluid inclusion brines, Dabusun Lake, Qaidam Basin, western China: Hydrology and paleoclimatology in arid environments. Palaeogeography Palaeoclimatology Palaeoecology. 1995;**117**:279-290

[60] Li MH, Yan MD, Wang ZR, Liu XM, Fang XM, Li J. The origins of the Mengye potash deposit in the Lanping-Simao Basin, Yunnan Province, Western China. Ore Geology Reviews. 2015;**69**:174-186

[61] Timofeeff MN, Lowenstein TK, Da Silva MAM, Harris NB. Secular variation in the major-ion chemistry of seawater: Evidence from fluid inclusions in cretaceous halites. Geochimica et Cosmochimica Acta. 2006;**70**(8):1977-1994

[62] Zhong XY, Yuan Q, Qin ZJ, Wei HC, Shan FS. The sporo-pollen analyses and ore-forming age of Nong Bok formation in Khammouane, Laos. Acta Geoscientica Sinica. 2012;**33**(3):323-330 (in Chinese with English abstract)

[63] Qu YH, Yuan PQ, Shuai KY, Zhang Y, Cai KQ, Jia SY, et al. Potash Forming Rules and Prospects of Lower Tertiary in Lanping-Simao Basin,

Yunnan. Beijing: Geological Press; 1998 (in Chinese with English abstract)

[64] Hasegawa H, Imsamut S, et al. Thailand was a desert' during the mid-Cretaceous: Equatorward shift of the subtropical high-pressure belt indicated by Eolian deposits (Phu Thok Formation) in the Khorat Basin, northeastern Thailand. Island Arc. 2010;**19**(4):605-621

[65] Utha-Aroon C. Continental origin of the Maha Sarakham evaporites, northeastern Thailand. Journal of Southeast Asian Earth Sciences. 1993;**8**(1-4):193-203

[66] Tan H, Ma H, Li BK, Zhang XY, Xiao YK. Strontium and boron isotopic constraint on the marine origin of the Khammouane potash deposits in southeastern Laos. Chinese Science Bulletin. 2010;**55**(27):3181-3188

[67] Zhang DW. Magnetostratigraphic studies of the Potash-bearing strata of the Lanping-Simao and the Vientiane Basins and their tectonic implications [thesis]. University of Chinese Academy of Sciences; 2016 (in Chinese with English abstract)

[68] Hansen BT, Wemmer K, Pawlig S, et al. Isotopic evidence for a Late Cretaceous age of the potash and rock salt deposit at Bamnet Narong, NE Thailand. In: Symposium on the Geology of Thailand, Bangkok; August 2002; Extended Abstract; 2002. pp. 26-31

[69] Hite RJ, Japakasetr T. Potash deposits of the khorat plateau, Thailand and Laos. Economic Geology. 1979;**74**(2):448-458

[70] Sun SR, Li MH, Yan MD, Fang XM, Zhang GX, Liu XM. et al. Bromine content and Br/Cl molar ratio of halite in a core from Laos: Implications for origin and environmental changes. Carbonates and Evaporites. 2019. DOI: 10.1007/s13146-019-00508-0

[71] Siemann MG. Extensive and rapid changes in seawater chemistry during the Phanerozoic: Evidence from Br contents in basal halite. Terra Nova. 2003;**15**(4):243-248

[72] Lu FH, Meyers WJ, Schoonen MA. S and O ($SO_4$) isotopes, simultaneous modeling, and environmental significance of the Nijar Messinian gypsum, Spain. Geochimica et Cosmochimica Acta. 2001;**65**(18):3081-3092

[73] Pierre C. Isotopic evidence for the dynamic redox cycle of dissolved sulphur compounds between free and interstitial solutions in marine salts pans. Chemical Geology. 1985;**53**:191-196

[74] Bottrell SH, Newton RJ. Reconstruction of changes in global sulfur cycling from marine sulfate isotopes. Earth Science Reviews. 2006;**75**:59-83

[75] Xu JX. Geochemistry and genesis of Mengyejing potash deposits [thesis]. Yunnan: Chinese Academy Science; 2008 (in Chinese with English abstract)

[76] Claypool GE, Holser WT, Kaplan IR, Sakai H, Zak I. The age curves of sulfur and oxygen isotopes in marine sulfate and their mutual interpretation. Chemical Geology. 1980;**28**:199-260

[77] Paytan A, Kastner M, Campbell D, Thiemens MH. Seawater sulfur isotope fluctuations in the Cretaceous. Science. 2004;**304**:1663-1665

[78] Hurtgen MT, Suits NS, Kaufman AJ. The sulfur isotopic composition of Neoproterozoic seawater sulfate: Implications for a snowball earth? Earth and Planetary Science Letters. 2002;**203**(1):413-429

[79] Hess J, Bender ML, Schilling JG. Evolution of ratio of strontium 87 to strontium 86 in seawater from

cretaceous to present. Science. 1986;**231**:979-984

[80] Krabbenhöft A, Eisenhauer A, Böhm F, Vollstaedt H, Fietzke J, Liebetrau V, et al. Constraining the marine strontium budget with natural strontium isotope fractionations (87Sr/86Sr, δ88/86Sr) of carbonates, hydrothermal solutions and river waters. Geochimica et Cosmochimica Acta. 2010;**74**:4097-4109

[81] Rahaman W, Singh SK, Sinha R, Tandon SK. Sr, C and O isotopes in carbonate nodules from the Ganga Plain: Evidence for recent abrupt rise in dissolved 87Sr/86Sr ratios of the Ganga. Chemical Geology. 2011;**285**:184-193

[82] Bo Y, Liu CL, Zhao YJ, Wang LC. Chemical and isotopic characteristics and origin of spring waters in the Lanping-Simao Basin, Yunnan, Southwestern China. Chemie der Erde-Geochemistry. 2015;**75**:287-300

[83] Li MH, Yan MD, Fang XM, Zhang ZJ, Wang ZR, Sun SR, et al. Origins of the Mid-Cretaceous evaporite deposits of the Sakhon Nakhon Basin in Laos: Evidence from the stable isotopes of halite. Journal of Geochemical Exploration. 2018;**184**:209-222

[84] Spivack AJ, Edmond JM. Boron isotope exchange between seawater and the oceanic-crust. Geochimica et Cosmochimica Acta. 1987;**51**(5):1033-1043

[85] Palmer MR, Helvaci C. The boron isotope geochemistry of the neogene borate deposits of western Turkey. Geochimica et Cosmochimica Acta. 1997;**61**(15):3161-3169

[86] Vengosh A, Starinsky A, Kolodny Y, Chivas AR, Raab M. Boron isotope variations during fractional evaporation of sea water: New constraints on the marine vs. nonmarine debate. Geology. 1992;**20**:799-802

[87] Pagani M, Lemarchand D, Spivack A, Gaillardet J. A critical evaluation of the boron isotope-pH proxy: The accuracy of ancient ocean pH estimates. Geochimica et Cosmochimica Acta. 2005;**69**(4):953-961

[88] Paris G, Gaillardet J, Louvat P. Geological evolution of seawater boron isotopic composition recorded in evaporites. Geology. 2010;**38**(11):1035-1038

[89] Foster GL, Pogge von Strandmann PAE, et al. Boron and magnesium isotopic composition of seawater. Geochemistry, Geophysics, Geosystems. 2010;**11**(8):Q08015

[90] Xiao Y, Sun D, Wang YH, Qi HP, Jin L. Boron isotopic compositions of brine, sediments, and source water in Da Qaidam Lake, Qinghai, China. Geochimica et Cosmochimica Acta. 1992;**56**:1561-1568

[91] Swihart GH, Moore PB. Boron isotopic composition of marine and nonmarine evaporite borates. Geochimica et Cosmochimica Acta. 1986;**50**(6):1297-1301

[92] Rose EF, Chaussidon M, France-Lanord C. Fractionation of boron isotopes during erosion processes: The example of Himalayan rivers. Geochimica et Cosmochimica Acta. 2000;**64**(3):397-408

[93] Lemarchand D, Gaillardet J, Lewin E. Boron isotope systematics in large rivers: Implications for the marine boron budget and paleo-pH reconstruction over the Cenozoic. Chemical Geology. 2002;**190**(1):123-140

[94] Liu WG, Xiao YK, Peng ZC, An ZS, He XX. Boron concentration and isotopic composition of halite from experiments and

salt lakes in the Qaidam Basin. Geochimica et Cosmochimica Acta. 2000;**64**(13):2177-2183

[95] Kloppmann W, Négrel P, Casanova J, Klinge H, Schelkes K, Guerrot C. Halite dissolution derived brines in the vicinity of a Permian salt dome (N German Basin). Evidence from boron, strontium, oxygen, and hydrogen isotopes. Geochimica et Cosmochimica Acta. 2001;**65**(22):4087-4101

[96] Zhang X, Ma H, Ma YQ, Tang QL, Yuan XL. Origin of the late Cretaceous potash-bearing evaporites in the Vientiane Basin of Laos: δ11B evidence from borates. Journal of Asian Earth Sciences. 2013;**62**(0):812-818

[97] Zhang CW, Gao DL, Zhang SY, Tang QL, Shi L. Comparison of geochemistry characteristics in Palaeocene salt-bearing strata of Lanping-Simao Basin and Chuxiong Basin. Journal of the Salt Lake Research. 2011;**19**(3):8-14 (in Chinese with English abstract)

[98] Lemarchand D, Gaillardet J, Lewin EÂ, AlleÅgre CJ. The influence of rivers on marine boron isotopes and implications for reconstructing past ocean pH. Nature. 2000;**408**:951-954

Chapter 4

# Mapping the Stable Isotopes to Understand the Geo-Structural Control of Groundwater Recharge and Flow Mechanisms (Case Study From the Northeastern Basin of the West Bank)

*Saed Khayat, Amer Marei and Zaher Barghouthi*

## Abstract

Conventional stable isotopic technique was used to differentiate between the potential recharge sources and mixing and flow mechanisms in the Northeastern basin of the West Bank. The isotopic signatures from deep wells show two main fingerprints with respect to recharge sources and mechanisms. These are wells located in the upper part of the Faria fault system and along the Rujeib Moncline which are fed by triggered water in-line the fault system in the south and deep wells surrounded by the Anabta anticline to the west which are fed by the exposed Jerusalem-Hebron formations. This suggests a mixing process with freshwater sources that mainly flow to the system from southern mountains. The isotopic signatures from the shallow well in Marj Sanoor wells and Nassariyeh in the upper Faria well suggest a kind of partial recharge from the Marj Sanoor Lake leaking to the upper Faria Graben area and participating in the recharge process of these wells. The whole finding out of this project might be used for tuning and revision of the groundwater model that has been built by the Palestinian Water Authority.

**Keywords:** isotope hydrology, Palestine, Northeastern aquifer, recharge mechanism, groundwater salinity

## 1. Introduction

Providing the Palestinian people with their water needs is the main concern for the Palestinian Water Authority, Ministry of Agriculture, as well as for water service providers. During the last 20 years, the annual average water consumption from the Northeastern aquifer reached 25 MCM [1]; this is due to improvement of water infrastructure including drilling new domestic deep wells, improvement of water institutions, and increase in the public awareness.

Tapping of groundwater, using spring water, purchase of water from Mekerot Israeli Company, and collection of rainwater are the sources for domestic and agricultural water in the West Bank. In this context, groundwater is the main one which covers about 70% of the total water supply which is about 100–145 MCM/a from the three water basins in the West Bank, namely, Eastern, Western, and Northeastern (the study area) [1]. During the last two decades, many domestic deep wells were drilled in the Mountain Aquifer in order to improve domestic water supply, where hundreds of illegal groundwater boreholes are drilled in the shallow aquifer systems mainly in Jenin and Jericho districts.

The Mountain Aquifer system with its three groundwater basins, namely, Western, Eastern, and Northeastern basins, covers most of the West Bank area (**Figure 1a**). The Western and Eastern basins cover the western and eastern parts of the West Bank and extend from Hebron in the south to Jerusalem and Ramallah in the north, while the Northeastern locates in the northern part of the West Bank, within the boundary of Nablus-Beit Qad syncline. About 410,000 Palestinians are living in three main districts (Nablus, Jenin, and Tubas), within the surface catchment area of this basin. These districts include large cities such as Nablus, Jenin, Tubas, and Tammon, beside many small municipalities, villages, and refugee camps [2].

Due to the fact that groundwater is the main source of water in the West Bank, management of this source is a high priority by line ministries, so solid and advance scientific knowledge is essential for sustainable management of the water resources [4]. Identification of recharge-discharge zones, groundwater flow regimes, connection between sub-basins, as well as shallow and deep aquifer systems are vectors for better management process. Applying environmental stable isotopic is a method in hydrogeology that is used to help in identification of these vectors, so we use $O^{18}_{[SMOW]}$ ‰ and D in the Northeastern Basin (NE Basin). We also combined the isotope analysis method, with the geological and hydrogeological setting of the sub-catchment areas [5]. High attention is given to the role of the main structural features of the groundwater flow regimes.

Recharge mechanisms, flow direction, and groundwater resident time are generally quite difficult to measure directly. Measurement of recharge flow can be exacerbated by preferential flow (i.e., macropore flow) in the unsaturated zone, although preferential flow paths are of greatest concern as potential conduits for rapid contamination of aquifers. The above factors, in addition to

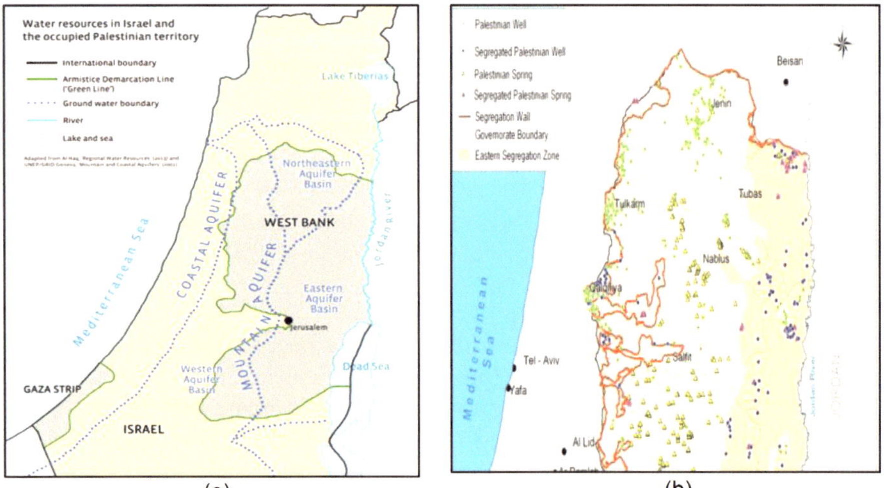


**Figure 1.**
*(a) Location of the Northeastern Basin and (b) urban zones in the NE basin catchment area [3].*

temporal and spatial variability, greatly complicate the estimation of basin-wide recharge rates and flow mechanisms. Estimation methods include use of water budgets, tracers, geophysics, and simulation models. Because of the inherent uncertainties in any method, it is often advisable to apply multiple techniques for any study.

Isotopes as tracer's isotopes for mapping the groundwater recharge and flow mechanisms are important tools in groundwater research and in sustainable management of groundwater resources. Important applications in shallow and deep groundwater include estimation of groundwater recharge and evaluation of the fate of contaminants, because meaningful groundwater deviation from the local meteoric water line LMWL gives the possibility to determine the residence time of groundwater and the dissolved contaminants [6–8].

## 2. Hydrogeological settings in NE Basin of the West Bank

Sedimentary rocks of the Upper Cretaceous to Quaternary ages cover most of the surface areas of the West Bank. Old rock formation exposed deep eroded area at the top of the anticlines or in deep wades and along some structural features (**Figure 2a**). The development of the structural features took place during the Late Upper Cretaceous-Tartary ages, where many structural features are still active. In the study area, two anticlines and one syncline in addition to the Faria Graben are main structural features. The Anabta anticline with northwest direction, parallel to the Faria anticline with northeast axis direction, borders the study area from the west and east, respectively, where the lines at the top of booths consider as surface/sub-surface water divide. Nablus-Beit Qad syncline with northeast axis direction locates between the two anticlines [9]. Due to the erosion process, the top of the anticlines is eroded, where rock formation of Cenomanian age crops out at the top of Faria and Anabta anticline, where carbonate rock of Eocene to Quaternary ages covers the central part of Nablus-Beit Qad syncline (**Figure 2b**).

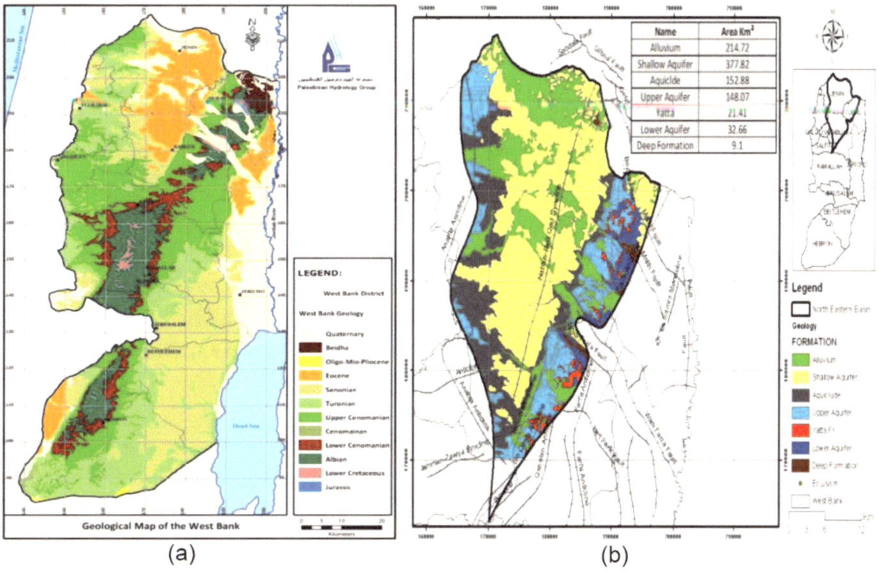


**Figure 2.**
*(a and b) Geological map of the West Bank and NE Basin.*

### 2.1 Stratigraphy

The stratigraphy sequences of the sedimentary rocks are the following from youngest to the oldest (**Figure 1b**) [10]:

#### *2.1.1 Alluvial deposits (Quaternary to recent age)*

It consists of alluvial deposits, mainly sand and gravel in flat and depression areas within the syncline. Brown earth Rendzina is the dominant soil type. The alluvial deposits cover an area of about 215 $km^2$ and normally overlay the Jenin subseries formation (**Figure 2b**).

#### *2.1.2 Jenin subseries (Eocene age)*

It consists of six members: relief limestone, nummulitic limestone, karstified limestone, limestone, and chalky limestone. The thickness of this subseries varies from one site to another depending on the location within the syncline but generally range between 100 and 300 m. The series is considered as a local shallow aquifer system that is used mostly in the agricultural sector. The majority of the springs in the NE Basin drain water from this aquifer; in addition to that, most of the groundwater boreholes (up to 350 m depth) in Jenin governorate are tapping water from this system. This rock formation covers an area of about 378 $km^2$ (**Figure 2b**). This formation overlays the Abu Dis formation in the central part of the syncline, where the Jerzim Group crop outs in the southern part of Nablus city.

#### *2.1.3 Jerzim group (Maastrichtian age)*

This group crops out in Jerzim Mountain 832 m above sea level within the southern part of the syncline to the south of Nablus city. It consists of chert nodules, chalk, and chalky limestone. The thickness is about 400 m. This group is considered as a local aquifer, where many springs in Nablus city drain water from this formation along the contact line between this layer and the underlying Abu Dis chalk unit.

#### *2.1.4 Abu Dis unit (Senonian age)*

It is composed of a massive thick hard chalk unit, interbedded with two bands of highly fractured cherty layers with a distance of 2–5 m between the two bands; the material of the upper part of unit l becomes soft and unclear in bedding. The unit exposes to the surface over the anticline flanks (**Figure 2b**) with a thickness range between 100 m at the edges and 500 m in the middle of the syncline. This unit is considered as an impermeable layer that separates the shallow Eocene aquifer from the underlying Jerusalem formation that is considered as the upper part of the Upper Mountain Aquifer system. The chalky units cover an area of about 153 $km^2$.

#### *2.1.5 Jerusalem formation (Turonian age)*

This formation consists of thin-bedded highly fractured limestone and dolomitic limestone. The lower part consists mainly of rosy limestone, where oyster fossil could be found at the top of this formation with variable thickness range between 70 and 150 m thick. This formation is cropped out mainly over the anticline and considered as recharge zone, where below the central part of the syncline consider as good aquifer. This formation overlays the Bethlehem formation (**Figure 2b**, Upper Aquifer).

### *2.1.6 Bethlehem formations (Upper Cenomanian age)*

This formation consists of 50–120 m of thin-bedded limestone and marly limestone which is highly karstified. Large caves and voids are common phenomena within this formation. This formation outcrop also covers the anticline flanks, and considers as the Jerusalem formation as good aquifer in the central and western part of the basin (**Figure 2b**, Upper Aquifer).

### *2.1.7 Hebron formations (Upper Cenomanian age)*

It consists of 105–250 m of thick bedded limestone and dolomite; it is highly fractured and karstified. This formation is cropped out also over the anticlines flanks and in deep eroded Wadis like Wadi Al Faria (**Figure 2b**), in these sites; it considers are recharge zone, where within the syncline consider as a target layer for groundwater subtraction.

### *2.1.8 Yatta formation (Lower Cenomanian age)*

It is composed mainly of marl and marly limestone; this formation is considered as an aquiclude in the southern part of the West Bank and separates the Upper from the Lower Aquifer system, but in the study area, this formation is more of limestone than marl and crops out at the flank of the anticline and is considered as part of the Upper Aquifer system.

The catchment area of the Upper Aquifer system (Jerusalem, Bethlehem, and Hebron formation) is about 148 $km^2$, where the area of Yatta formation is about 21 $km^2$. Together, Jerusalem, Bethlehem, and Hebron formation build up the Upper Aquifer system of the Mountain Aquifer system and are separated from the Lower Aquifer system through impermeable marl layer of Yatta formation of Lower Cenomanian age. Older formations such as Upper and Lower Beit Kahel do not out crop in the study area.

### *2.1.9 Upper Bet Kahel formation (Albian)*

It consists of a 160–190-m-thick well-bedded limestone and dolomite. The lower part of this formation is made up of limestone with thin layers of porous dolomite interchanging with marly limestone and calcite massive limestone near the base.

### *2.1.10 Lower Bet Kahel Formation (Albian)*

It consists of gray limestone layers alternating with layers of shale and marl in the lower part, whereas the upper part is made up of gray to brown dolomite with clayey and marly limestone.

Both formations are considered as Lower Aquifer system of the Mountain Aquifer in the southern part of the West Bank, but in the northern part of the Upper and Lower Aquifer system, they are considered as one hydrological system. The outcrop area of the deep aquifer system is about 33 $km^2$. The two formations crop out also in deep eroded streams within the Faria anticline as well as within the Faria Graben.

## 2.2 Hydrology

Rainy months extend from October to May, where 70% of the rainfall takes place between December and February. **Figure 3** shows the rain fall distribution during

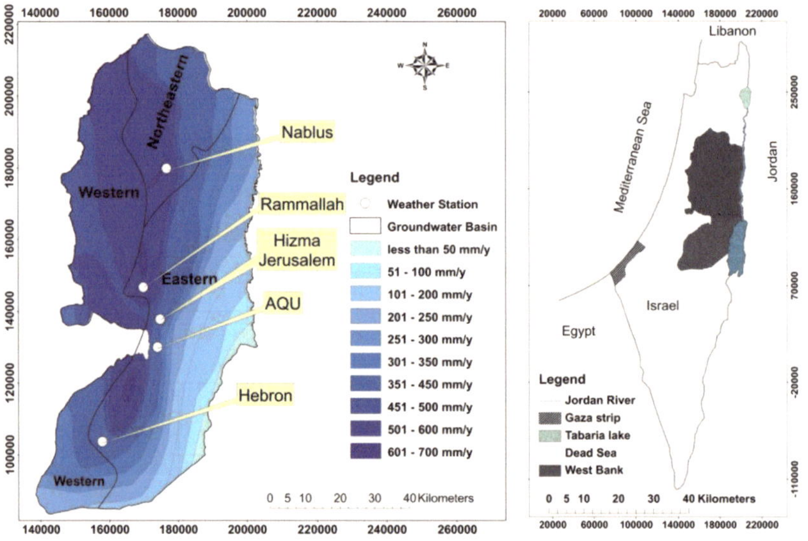


**Figure 3.**
*Actual average rainfall in the West Bank in the hydrological year 2010/2011.*

the hydrological year 2015/2016, where three high rainfall zones are identified namely within the boundary of the basin, these are north of Nablus, Selet al Thaher with 600 and 800 mm/a respectively [11]. It's also noticed that rainfall decreases in the eastward direction of the Faria Graben which locates more within the rainfall shadow site (**Figure 3**). The average monthly temperature during December, January, and February is 11, 14, 17°C. respectively; this indicate that the losses of water through evapotranspiration is relatively low during these months which improve the groundwater recharge rate [12].

## 2.3 Groundwater aquifer systems

The Northeastern Basin covers 959 km$^2$ of surface area (**Figure 1b**), depending on surface water shed divide; within the basin, two aquifer systems are identified; Mountain Aquifer with rock layers related to the Upper Cretaceous age, and shallow aquifer with rock layers related to the Tertiary-Quaternary eras (Eocene-Miocene age). Both systems are hydraulically separated from each other in most of the basin especially in the central part but seems to be connected where deep structural features strike the rock layers of both systems, such as in Al Faria Graben [13].

Recharge process for both aquifer systems takes place wherever the rock formation is outcropped and exposed directly to the rainfall or underlying thin soil layers. Marei et al. estimate the groundwater recharge rate, by using chloride mass balance method for the study area, of about 95.2 and 269.7 mm/year, with a total average recharge volume of 138.5 MCM/year (**Figure 4**) [14], while the total calculated recharge rate by the previous study of the authors is 107.1 MCM/a [14]. Recharge rate can be higher than estimated when karstic and high fractured rock layers are cropped out at the surface such as the formation of the Upper and Lower Mountain aquifer system at the two anticline flanks in the west and in the east, in the other hand ground recharge decrease to about zero from Abu Dis formation "Chalky Unit". The formation of shallow aquifer system is exposed mainly in the central part of the basin, and water body responds quickly to rainfall.

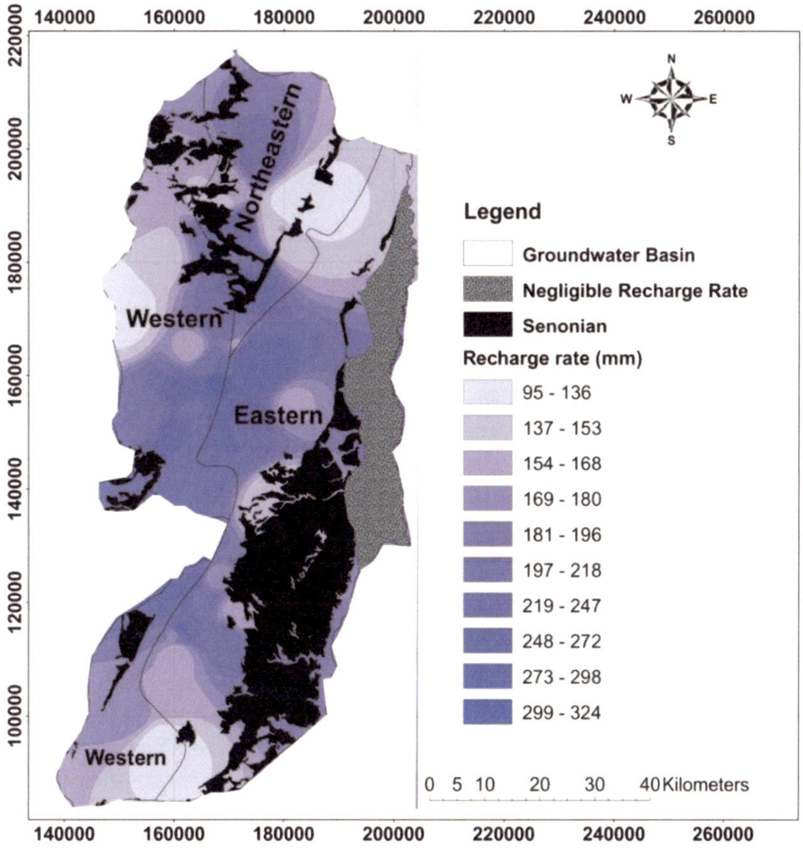


**Figure 4.**
*Recharge rate over the West Bank including the NE Basin (Marei et al., 2011).*

| | Mountain Aquifer "Cretaceous age" | | Shallow Aquifer "Eocene" |
|---|---|---|---|
| | Eastern flank of "Anabta anticline" | Western flank of "Al Faria anticline" | Nablus-Beit Qad "syncline" |
| Outcrop area with $km^2$ | 60 | 95 | 378 |
| Recharge rate | 210 mm/year | 210 mm/year | 200 mm/year |
| Recharge volume | 12.6 MCM | 19.9 MCM | 75.6 MCM |
| Total recharge in MCM/a | 32.5 MCM | | 75.6 MCM |

**Table 1.**
*Recharge volume of NE Basin.*

**Table 1** summarized the recharge volume of the deep and shallow aquifer systems depending on the chloride mass balance method [14].

## 2.4 Groundwater flow regimes

Two main groundwater flow regimes are assumed to present in the study area. These are as follows: a SW-NE main groundwater flow direction parallel to

Nablus-Beit Qad syncline axis with historical discharge site in Hiteen and Ein Jaloot spring and a flow direction from both anticline flanks (Anabtaa, and Al Faria) to the center of the syncline which joins the NE flow direction; these flow directions take place within the Mountain Aquifer system [15, 16]. Addition flow direction to the southeast is governed through Al Faria fault system "Graben" that diverted groundwater to flow in this direction [17].

## 3. Methodology

Integrated isotopic tools were used to investigate the effect of complex geologic structure on the groundwater residence times and respective potential sources, mixing, and recharge mechanisms [18]. In order to achieve the abovementioned objectives, 82 groundwater samples were obtained from different areas in the Northeastern basins. The samples represent 8 springs, 20 wells from shallow Eocene aquifer in the plain zone of the study area, 7 wells in Sanoor swamp area, 10 shallow wells in Nassariyeh area in the upper part of Wadi Faria stream, 20 wells from the lower part of Faria stream, and 17 deep aquifer wells within and near the flanks of the NE basin (**Figure 5**). All samples were taken in the hydrological year 2017/2018. Several rainwater samples were obtained from rain gauges' stations that were constructed on the roofs of some schools all over the study area.

Groundwater samples for deuterium and $\delta^{18}O$ isotopes have been taken from the mentioned wells and spring. Samples for deuterium and $\delta^{18}O$ isotopes were collected with 25 ml bottles and sent to the Al-Quds University research lab for analysis. Samples were analyzed using laser spectroscopy for deuterium and $\delta^{18}O$ in ‰ in respect to Vienna Standard Mean Ocean Water (V-SMOW) standard; the precision of $\delta^{18}O_{[SMOW]}$ ‰ measurements is ±0.1‰; the precision of $\delta D$ values is ±2‰ [19].

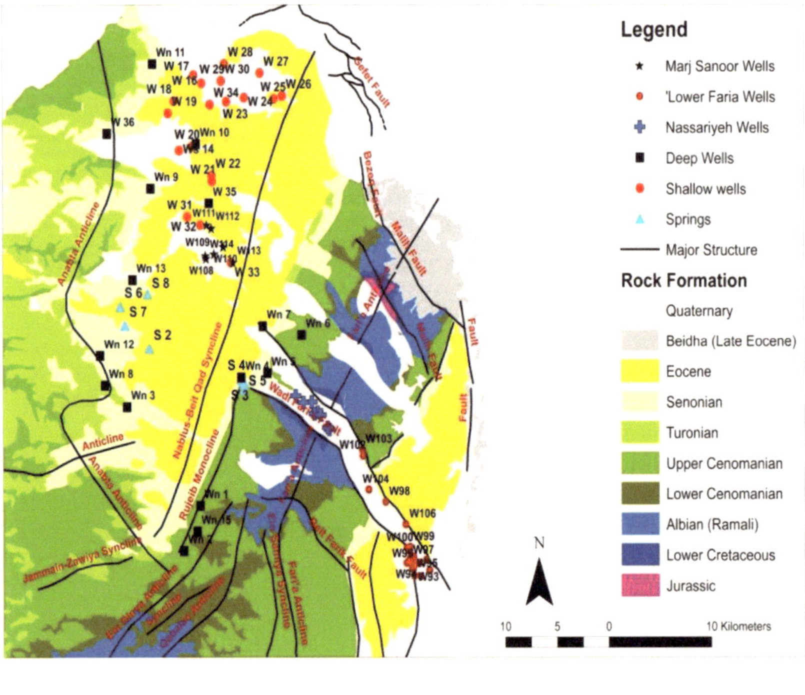


**Figure 5.**
*The study area including all the sampled wells and springs, including rock formations and structure.*

*DOI: http://dx.doi.org/10.5772/intechopen.90449*

## 4. Results and discussion

### 4.1 Results overview

The local meteoric water line shows the same slope for the Mediterranean Meteoric Water Line but with more enriched deuterium excess. This might refer to the formation of a large swamp lake in Sanoor area which resulted from the inundation from the runoff drained to the area from the surrounded mountains causing high humid conditions in the area (**Figure 6**).

The data show a wide range of isotopic signatures, which reflects wide variations with respect to recharge mechanisms and groundwater flow directions.

**Figure 5** shows the distribution of sampled wells and different geological structures that control the hydrology of the region. As it is mentioned above, the structural geology is highly controlling the hydrological flow system in the region. The main structure that might play an important role in this regard is the Faria fault system which might control the groundwater flow regime in the eastern part of the NE Basin.

Spring systems in both locations (Bathan in the east and Nablus in the north west) show closed signatures to the local meteorological line, which indicate rapid freshwater input; the other end member of shallow wells within the middle of the syncline shows the most enriched signatures (**Figure 6**).

Other wells, which show signatures in between, have different recharge mechanisms which need to be separated in details with respect to isotopic signature from each group.

The following sections illustrate the relations between different aquifers as well as the recharge mechanism for each system.

### 4.2 Deep wells

The isotopic signatures from deep wells show two main fingerprints with respect to recharge sources and mechanisms.

First, the deep wells that are located within the area of the Faria fault system, southern part of the syncline and upper part of Faria fault system, show depleted signatures that are more or less closed to springs and LMWL, while other deep wells reflect high variation in isotopic enrichment with respect to its depth and location

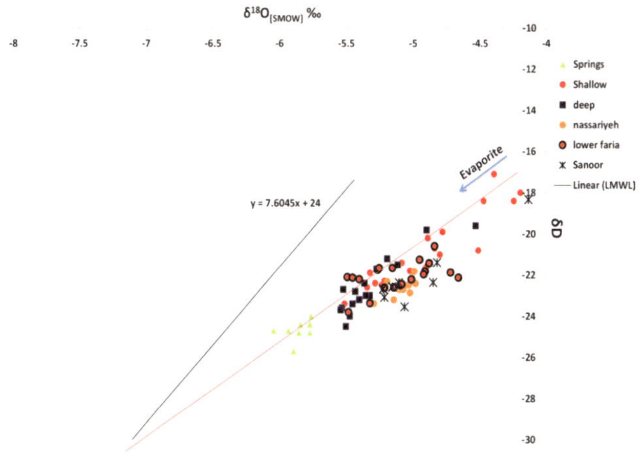


**Figure 6.**
*Deuterium vs. $\delta^{18}O_{[SMOW]}$ ‰ for the whole wells and springs in the study area.*

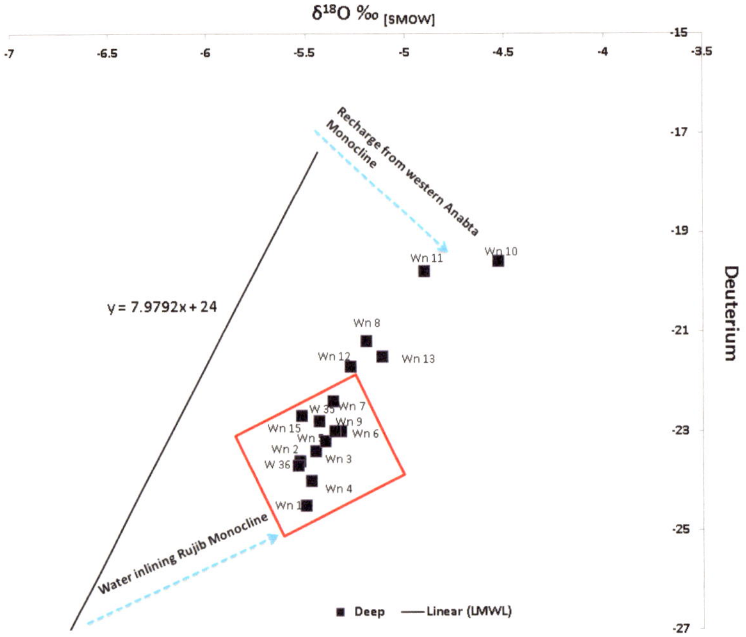


**Figure 7.**
*Relation between $\delta^{18}O_{[SMOW]}$ ‰ and deuterium for the groundwater from deep aquifer shows different recharge mechanisms for each cluster.*

(**Figures 6** and 7). Some of the deep wells show obvious close relation to the recharge that feeds the aquifer layers through the exposed Jerusalem-Hebron formations on Anabta anticline (**Figure** 7), where the wells within this area show more enriched $\delta^{18}O$ signatures than those near the upper Faria fault system (**Figure** 7).

The isotopic signatures of deuterium show clear differences between each deep well cluster with an average shifting of 4‰. These differences are more or less related to the recharge locations with different altitudes [20].

The deep wells in the south and upper part of Faria show relatively more depleted deuterium than those deep wells that receive direct recharge from the western Anabta anticline outcrops. The elevation of Anabta anticline in the western part of the basin has an average altitude of 300 m above sea level, while the southern elevation over the mountains in Nablus and Salfit areas to the south, from where the recharge for deep Faria well cluster is expected, reaches an average of 500 m above sea level.

However, the $\delta^{18}O_{[SMOW]}$ ‰ signatures show slight shifting between both clusters, with more slight enrichment for the wells near to the western Anabta anticline that reach around −1‰. This also reflect different recharge mechanisms and different hydrological flow conditions from each source [21].

**Figure 8** shows the suggested recharge zones and flow mechanisms for each of deep well cluster.

### 4.3 Shallow wells

This group of wells can be divided into three major categories according to its locations: shallow Eocene wells that are distributed in the plain area of syncline, Sanoor wells which are belonging to the same previous area but located directly within the area of surface water swamp, and upper Faria (Bathan) shallow wells in Nassariyeh area and lower Faria shallow wells to the southeast.

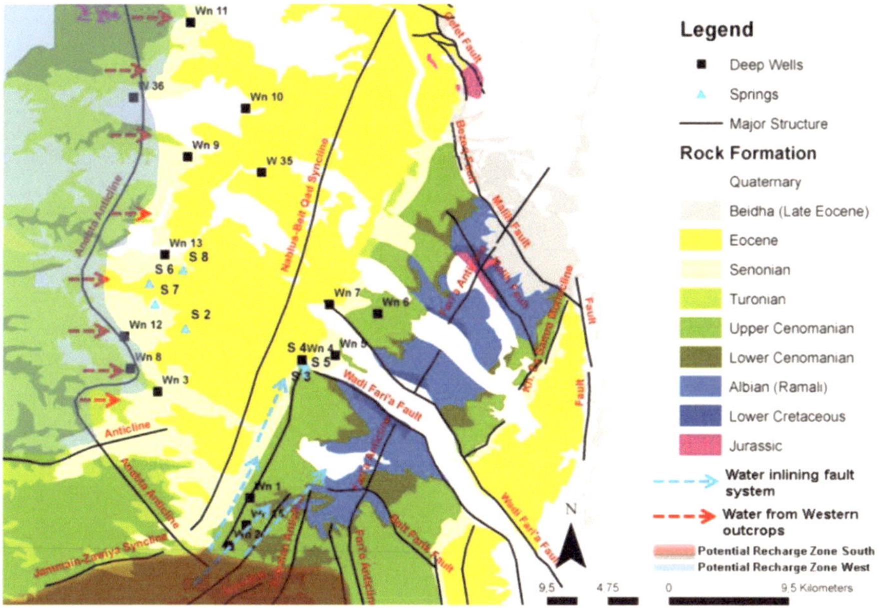


**Figure 8.**
*Suggested recharge zones and flow mechanisms for each of deep well cluster.*

According to isotopic signatures from these shallow wells, different recharge mechanisms for each group can be indicated. Also the isotope data reflect some hydrological connections between some groups. The hydrological relations as well as recharge mechanisms can be described with respect to each group.

### *4.3.1 Shallow wells within the Eocene plain area*

This includes the shallow wells that are dug in the Eocene and Quaternary alluvial areas in the northwest of Nablus-Bet Qad syncline (**Figure 5**). Those wells show high evaporation inputs with relatively high TDS content. The $\delta^{18}O_{[SMOW]}$ ‰ and deuterium values for the shallow wells in the Al Jalameh (north) indicate relatively enrichment deviation from the LMWL due to fractionation with the thick soil layer during slow infiltration. For these wells, the $\delta^{18}O_{[SMOW]}$ ‰ values reach −3.5‰ (**Figure 6**). The deviation from LMWL with the slope of 4 indicates an evaporation trend that increases toward the north of the study area where those wells tapped their water from (**Figure 6**).

In general, the problem of water deterioration in this group seems to be connected with the heavy abstraction rate from these wells. The slow replenishment, with such heavy abstraction, increases the evaporite salinity problem [22].

In general, Marj Sanoor wells show relatively enriched $\delta^{18}O$ signatures but less than the rest of shallow wells in the north (**Figure 6**). This might be due to the fact that the aquifer is located beneath the water swamp that is collected in the winter time and infiltrated slowly to the aquifer layers. The integration of the results with the results of other locations shows a connection between the infiltrated surface water from this group with some wells to the southeast as it will be described below.

### *4.3.2 Shallow wells in Nassariyeh*

The shallow wells in Nassariyeh that are located at the beginning of Faria structural faults show the same stable isotopic signatures as Marj Sanoor wells. This

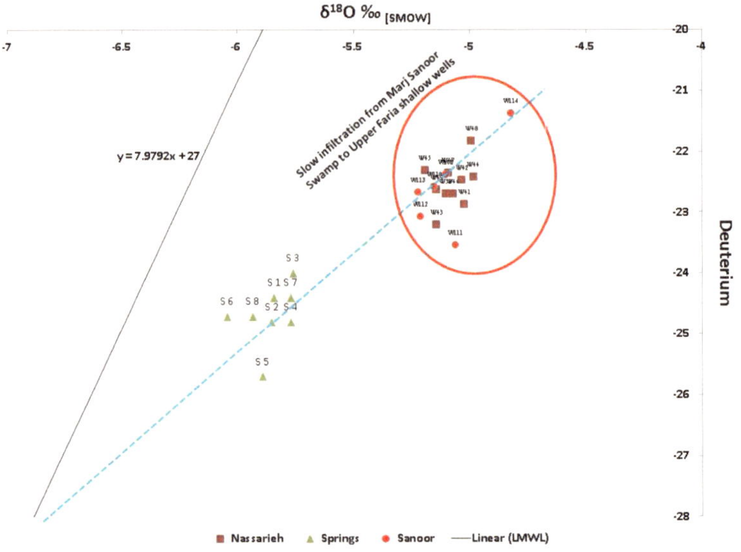


**Figure 9.**

*Relation between $\delta^{18}O_{[SMOW]}$ ‰ and deuterium for the groundwater from Nassariyeh and Sanoor shallow wells.*

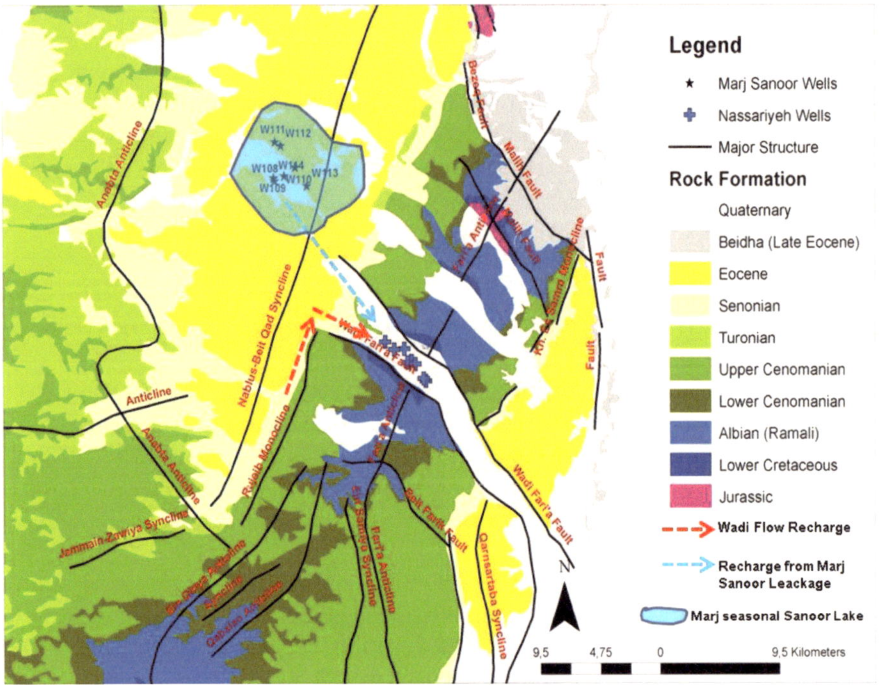


**Figure 10.**

*Suggested groundwater model for the water leakage from Marj Sanoor Lake to the Nassaria and upper Faria wells.*

strongly suggests a connection between seeping water from the seasonal Marj Sanoor Lake, which forms by the collected runoff from surrounding mountains in the late winter season to the wells that are located within the upper Faria and Bathan area (**Figure 9**).

The isotopes signatures suggest that the recharge mechanism for these wells is a mixing between water seepage from Marj Sanoor surface water and fresh water that

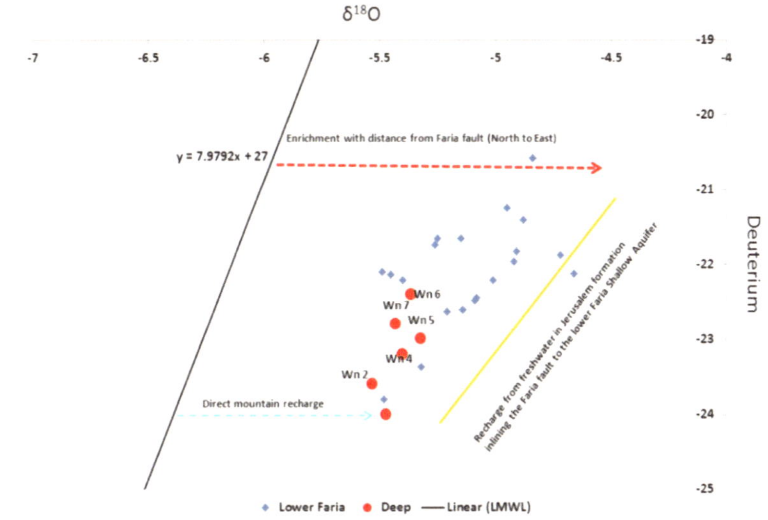


**Figure 11.**
*Relation between $\delta^{18}O_{[SMOW]}$ ‰ and deuterium for the groundwater from deep wells and lower Faria shallow wells.*

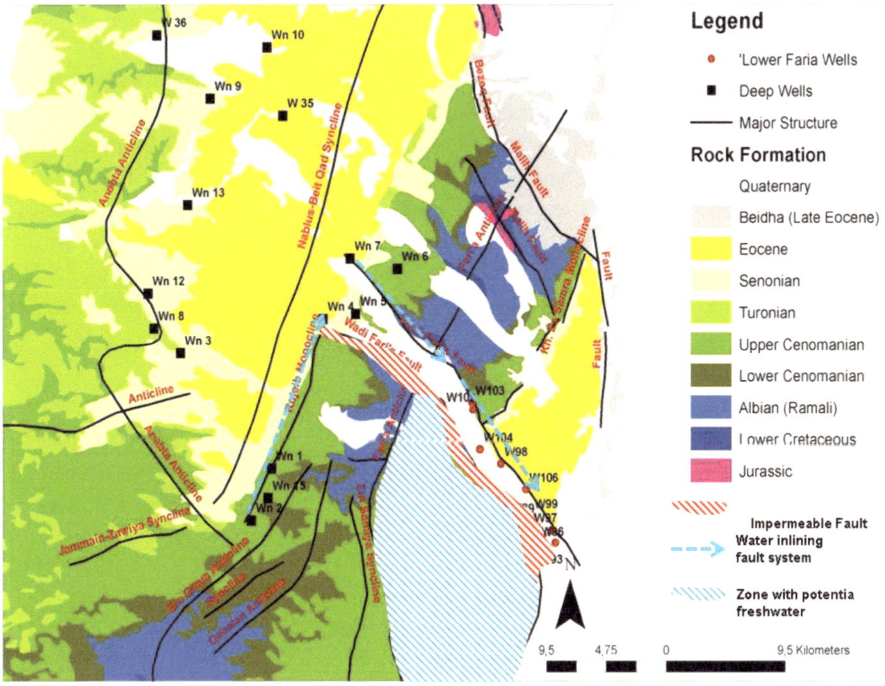


**Figure 12.**
*Model of groundwater recharge and flow mechanisms for the lower Faria wells from different sources.*

inline the Faria Fault that triggered from deeper Jerusalem formation, and seeping along the area of Faria Graben (**Figure 10**).

This finding can be used to efficiently utilize the surface water in the syncline area to artificially feed the wells further to the east, keeping the groundwater level in good standing all over the summer season. On the other hand, heavy abstraction from the shallow and deep wells within the syncline area might affect the productivity of Bathan and upper Faria wells.

### 4.3.3 Lower Faria shallow wells

The isotopic signatures from lower Faria shallow wells suggest strong correlation with the recharge of the deep well in the upper Faria part. Most of the shallow lower Faria wells show the same $\delta^{18}O$ and deuterium signatures for the deep wells of Bathan, Faria, Tubas, and Tammoun. This similarity emphasizes the unity of recharge mechanism for both locations which mainly come from Jerusalem formation of Turonian age, that triggered along the northern Faria fault and seeping to the wells drilled within lower Faria plain (**Figures 11** and **12**).

However, the isotopic signatures show enrichment trend with respect to distance from the fault to the middle and the south of the Wadi (**Figure 13**). This emphasizes that the main recharge source for the wells in the lower Faria is coming mainly from the northwest, in-line the Fault system (**Figure 12**).

This also can be an indicator about the limitation of the water recharge from the southern part of the Fault, which suggests in role that most of the recharge in the southern area is drain surfacely and sub-surfacely to the area constrains between the southern Faria Fault and Bet Forik Fault, where the mentioned area must be a good potential for freshwater production with sufficiently high amount.

## 5. Conclusion and recommendations

The isotopic signatures from deep wells show two main fingerprints with respect to recharge sources and mechanisms. Those are wells located in the upper part of Faria fault system and along the Rujeib Moncline which are fed by triggered water in-line the Fault system in the south and deep wells surrounded by Anabta anticline to the West which are fed by the exposed Jerusalem-Hebron formations. This suggests a mixing process with freshwater sources that mainly flow to the system from southern mountains. However, the impermeability of the southern part of Faria fault system makes this water diverted to the area constrain between the southern

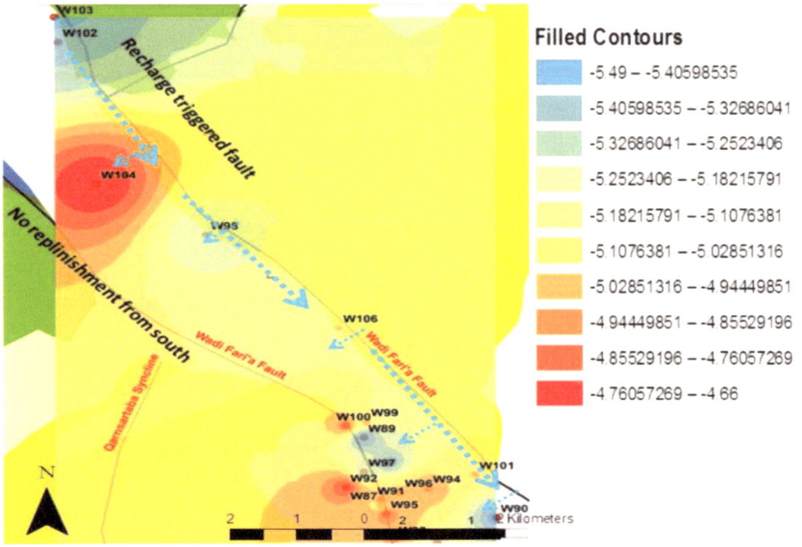


**Figure 13.**
*Spatial distribution of $\delta^{18}O_{[SMOW]}$ ‰ for the shallow wells in lower Faria shows relatively depleted signatures along the fault and more enriched to the center of the Graben.*

Faria fault and Bet Forik faults, where the mentioned area must be a good potential for freshwater production with sufficiently high amount (**Figure 13**).

The isotopic signatures from the shallow well in Marj Sanoor wells and Nassariyeh in the upper Faria well suggest a kind of partial recharge from the Marj Sanoor Lake that leaks to the upper Faria Graben area and participates in the recharge process of these wells. This finding can be used to efficiently utilize the surface water in the syncline area to artificially feed the wells further to the east, keeping the groundwater level in good standing all over the summer season. On the other hand, heavy abstraction from the shallow and deep wells within the syncline area might affect the productivity of Bathan and upper Faria wells.

The northern part of the Faria Graben fault which shows a good ability for freshwater transmission from different sources is extended further to the north-west, reaching the syncline area. The area of fault extension can be a good potential source for drilling new wells in the future.

The whole finding of this project might be used for tuning and revision of the groundwater model that has been built by the Palestinian Water Authority. The suggested new flow mechanisms and potential recharge zones can help the Palestinian stakeholders in good planning for the whole Northeastern aquifer system.

## Acknowledgements

We thank the International Atomic Agency for funding this work through the national project PAL7005. Special thanks should go to Gerald Reys, the Section Head of Asia and the Pacific in the Department of Technical Cooperation, for her follow-up and continuous administrative and technical support. Also many thanks go to Mr. Ismael Hroub, the NLO in the Palestinian Ministry of Health for his follow-up and administration support.

## Author details

Saed Khayat[1*], Amer Marei[2] and Zaher Barghouthi[3]

1 Palestine Technical University/Kadoorie (PTUK), Tulkarm, Palestine

2 Al-Quds University, East Jerusalem, Palestine

3 NARC—National Agricultural Research Centre, Jenin, Palestine

*Address all correspondence to: saed.khayat@gmail.com

## References

[1] Authority-PWA PW. Annual Water Status Report. Yearly Report, 2017. 2018

[2] Statistics-PCBS PCBo. Preliminary Estimates of Quarterly National Accounts (First Quarter 2018). 2018

[3] Arij A. Status of the Environment in the Occupied Palestinian Territory. Jerusalem Applied Research Institute (Arij), 2007.

[4] Abu Zahra BAA. Water crisis in Palestine. Desalination. 2001;**136**(1-3):93-99

[5] Sidle WC. Environmental isotopes for resolution of hydrology problems. Environmental Monitoring and Assessment. 1998;**52**(3):389-410

[6] Böhlke J, Denver J. Combined use of groundwater dating, chemical, and isotopic analyses to resolve the history and fate of nitrate contamination in two agricultural watersheds, Atlantic coastal plain, Maryland. Water Resources Research. 1995;**31**(9):2319-2339

[7] Clark I, Fritz P. Environmental Isotopes in Hydrogeology. New York: CRC Lewis Publishers, Inc; 1997. p. 328

[8] Cook P, Herczeg AL. Environmental Tracers in Subsurface Hydrology. Norwell: Kluwer Academic Publishers; 2000. p. 529

[9] Rosenthal E et al. Late cretaceous-early tertiary development of depositional basins in Samaria as a reflection of eastern Mediterranean tectonic evolution. AAPG Bulletin. 2000;**84**(7):997-1014

[10] Rofe and Raffety, West Bank. Hashemite Kingdom of Jordan. Central Water Authority. London: Rofe and Raffety Consulting Engineers Westminster; 1963

[11] Gunkel A et al. Model signatures and aridity indices enhance the accuracy of water balance estimations in a data-scarce Eastern Mediterranean catchment. Journal of Hydrology: Regional Studies. 2015;**4**(Part B): 487-501

[12] Saleh YFK. Artificial Groundwater Recharge in Faria Catchment A Hydrogeological Study, in Faculty of Graduate Studies. Faculty of Graduate Studies, An-Najah National University: 2009

[13] Rofe & Raffety 1965. West Bank hydrology 1963-1965: Analysis. Rofe & Rafferty Consulting Engineers Ltd., A., Report for Central Water Authority, Amman, Hashemite Kingdom of Jordan., West Bank hydrology 1963-1965. Report for Central Water Authority. Amman, Hashemite Kingdom of Jordan; 1965

[14] Marei A et al. Estimating groundwater recharge using the chloride mass-balance method in the West Bank, Palestine. Hydrological Sciences Journal-Journal Des Sciences Hydrologiques. 2010;**55**(5):780-791

[15] Najem AAI. Modeling Nitrate Contamination of the Eocene Aquifer, Palestine. An-Najah National University; 2008

[16] Rosenthal E, Meiler M, Flexer A. Structure-controlled groundwater flow and salinization paths in the Bet She'an and Harod Valleys, Israel. Environmental Earth Sciences. 2012;**66**(4):1071-1082

[17] Tubeileh HMSR. Groundwater Flow Modeling-Case Study of the EoceneAquifer in the West Bank, Palestine. 2003

[18] Gat JR. Oxygen and hydrogen isotopes in the hydrologic cycle. Annual

Review of Earth and Planetary Sciences. 1996;**24**:225-262

[19] Koziet J. Isotope ratio mass spectrometric method for the on-line determination of oxygen-18 in organic matter. Journal of Mass Spectrometry. 1997;**32**(1):103-108

[20] Araguás-Araguás L, Froehlich K, Rozanski K. Deuterium and oxygen-18 isotope composition of precipitation and atmospheric moisture. Hydrological Processes. 2000;**14**(8):1341-1355

[21] Barnes CJ, Allison GB. Tracing of water movement in the unsaturated zone using stable isotopes of hydrogen and oxygen. Journal of Hydrology. 1988;**100**(1-3, 143):-176

[22] Khayat S et al. Hydrochemical investigation of water from the Pleistocene wells and springs, Jericho area, Palestine. Hydrogeology Journal. 2006;**14**(1-2):192-202

Chapter 5

# Could Coral Skeleton Oxygen Isotopic Fractionation be Controlled by Biology?

*Anne Juillet-Leclerc*

## Abstract

During 1970s, coral skeleton oxygen isotope composition ($\delta^{18}O$) was regarded as the isotopic thermometer following thermodynamic rules. Recently, coral aragonite oxygen isotopic fractionation could appear to be controlled by biology, its rate being accelerated by an enzyme (carbonic anhydrase or CA). Such a new concept results of an original approach involving coral culture in controlled conditions. Environmental factors, temperature and also light have been tested on macrosize scale samples (some mg), and $\delta^{18}O$ revealed vital effects, anomalies compared with chemical and isotopic equilibrium, related to metabolic activity. $\delta^{18}O$ analyses at microsize scale (some μm), using ion microprobe, could display the mechanism of crystallisation, $\delta^{18}O$ fractionation responding to biological kinetic effects. The understanding of coral aragonite $\delta^{18}O$ is the absolute prerequisite to develop the first model of a coral proxy.

**Keywords:** oxygen isotopic fractionation, coral skeleton, culture, controlled conditions, temperature, light, organic matrix

## 1. Introduction

Coral colonies built the most important bioconstruction made of calcium carbonate ($CaCO_3$) of the world, with a calcification of about 2–6 $kg_{CaCO3}$ $m^{-2}$ $year^{-1}$ covering an area of about 284,300 $km^2$ [1]. This construction built from Jurassic results from the work of multiple small colonial organisms. The reefs, the biotic mound structure essentially made of corals as the Great Barrier Reef in Australia, are of major importance for marine ecosystems and biodiversity because they are the most productive and they host almost a third of all world fishes.

Corals are marine animals forming an aragonite (a polymorph of $CaCO_3$) skeleton. They are developed in two distinct ecosystems, essentially zooxanthellate corals or symbiotic ones living in shallow water and solitary colonies or integrated in elaborate reef framework in deeper depth than 50 until 2000 m. More than 793 coral species are spread over marine tropical zone [2]. Branched corals *Acropora* and massive corals *Porites* are ubiquitous genera [3]. We restricted this study to zooxanthellate corals.

Epstein [4, 5] demonstrated that skeletal carbonates of marine shells display similar oxygen isotopic composition ($\delta^{18}O_{CaCO3}$) relationship versus SST than inorganic calcium carbonate ($CaCO_3$) deposited from seawater at the same temperature, following thermodynamic laws.



IntechOpen

$$\delta^{18}O = \left\{ \left[ \left( ^{18}O/^{16}O \right)_{sample} / \left( ^{18}O/^{16}O \right)_{standard} \right] - 1 \right\} * 10^{-3}$$

The relationship was expressed as:

$$SST°C = 16.5–4.3 \left( \delta^{18}O_{CaCO3} - \delta^{18}O_{seawater} \right) + 0.14 \left( \delta^{18}O_{CaCO3} - \delta^{18}O_{seawater} \right)^2 \quad (1)$$

from [5] with sea surface temperature (SST) being sea surface temperature and $\delta^{18}O_{seawater}$ seawater isotopic composition. The authors underlined that coral skeleton presented poor interest [4].

However, after preliminary studies [6], Weber and Woodhead deduced, despite apparent isotopic disequilibrium between coral skeletal carbonate and ambient seawater, that $\delta^{18}O_{coral}$ was temperature dependent. To support this assumption [2], the authors conducted isotopic analyses of coral skeleton collected over wide range of temperatures. This data series still constitutes the most exhaustive oxygen isotopic database existing for corals. Weber and Woodhead concluded that the calibration between annual $\delta^{18}O$ and annual SST differed following each coral genus, and the isotopic disequilibrium was attributed to vital effect, anomalies compared with chemical and isotopic equilibrium, related to metabolic activity. Several models of mineralisation were proposed to explain the geochemical specificities of coral skeletons based on kinetic fractionation [7–9] disturbed by "vital effects". Other models, based on precipitation efficiency [10] or Rayleigh fractionation [11–13], were suggested.

We developed drastically different approach considering that corals are animals living in symbiosis with algae, building aragonitic skeleton intimately related to biological activity. In collaboration with biologists from CSM (Centre Scientific de Monaco), Stéphanie Reynaud and Christine Ferrier-Pagès, and Claire Rollion-Bard geochemist from IPGP (Institut Physique du Globe de Paris), we developed an innovative strategy on cultured *Acropora* to identify what was hidden under the term of vital effect of coral skeleton and to highlight the isotopic fractionation involved in. We focused our study on the stable oxygen isotopic ratio $\delta^{18}O$. Branched coral *Acropora* and massive ones *Porites* belong to different genera but responses to environmental forcing in terms of biological parameters and isotopic signatures are regarded as similar.

Our demonstration is structured as followed: first, the main coral features are highlighted; second, we describe coral culture proceeding; third, temperature and light test results are presented at microscopic size scale; and finally, we display the stable oxygen isotopic ratio $\delta^{18}O$ as indicator of deposit mechanism.

## 2. Main coral features

Shallow corals, because they are leaving in symbiosis with micro algae need light to benefit from photosynthetic activity.

### 2.1 Notions of coral morphology and biological activity

Coral skeleton is extracellular, located at the base of coral tissue, constituted of similar units, the polyps. Each polyp looks like a bag made by two layers of cells (**Figure 1**). Polyps are linked together by the coenosarc. Most of zooxanthellae are located within an internal layer (**Figure 1**).

Biological activity might be quantified. Photosynthesis and respiration were measured using the respirometry technique, which measured the changes in oxygen concentration at different light levels. Rates of net photosynthesis and respiration

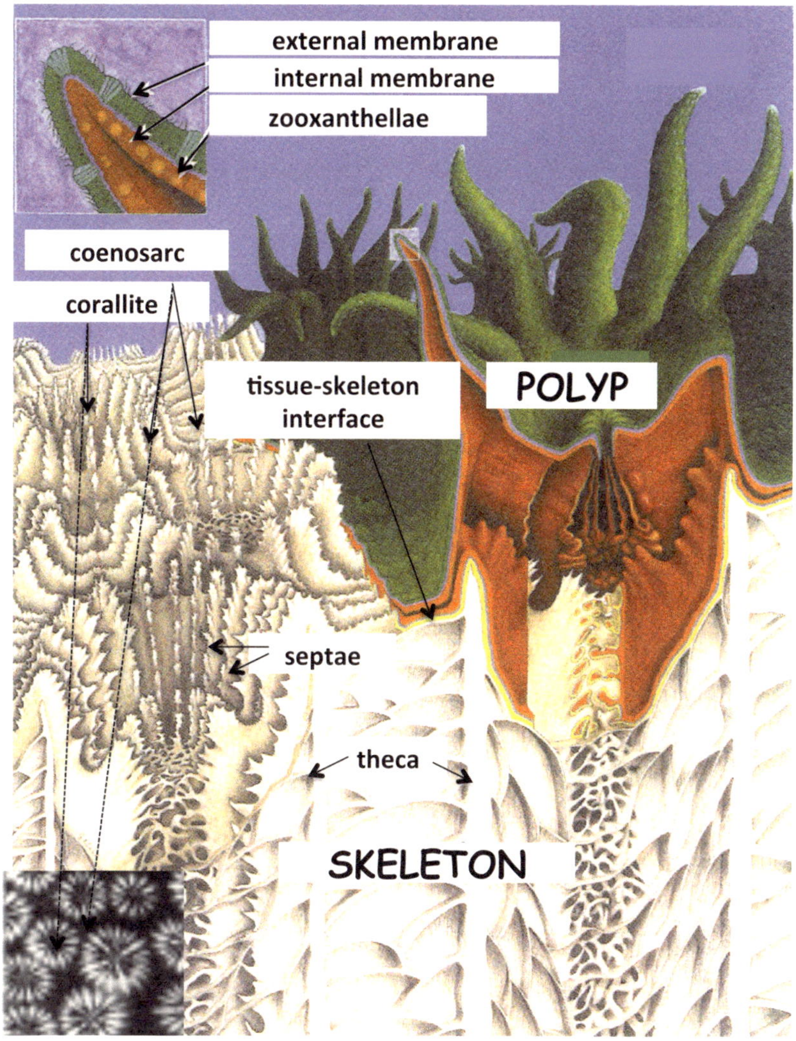


**Figure 1.**
*Organisation of the coral adapted from [14]. The colour polyp is the living organism, an animal building the white skeleton. The animal lives in symbiosis with algae, the zooxanthellae, located in the internal layer (the left side high corner). A detail of mesoscale skeletal architecture figures on the left side bottom corner.*

were estimated using a linear regression of $O_2$ against time [15]. By using two light intensities, at three temperatures (**Figure 2**), it appeared that increasing temperature enhanced photosynthetic activity, the effect arising with light intensity (**Figure 2a**) [15]. Additionally, coral growth rates might be estimated. Corals were weighed regularly using the buoyant weight technique and the surface expansion of the new skeleton formed was estimated [15]. It was generally accepted that calcification was light-enhanced (LEC or light enhanced calcification) during the day [16]. Zooxanthellae density and pigment concentration were determined under the light microscope using a counting chamber [15]. Coral symbiont distribution was not homogeneous on the skeleton and it differed following different coral genera, different depths in the fields. For example, tips of coral branches or other exposed surfaces were sun-adapted while most of the lowest parts were shade-adapted [15]. Iluz and Dubinsky [17] listed all the strategies developed by coral to optimise the light impinging on the zooxanthellae.

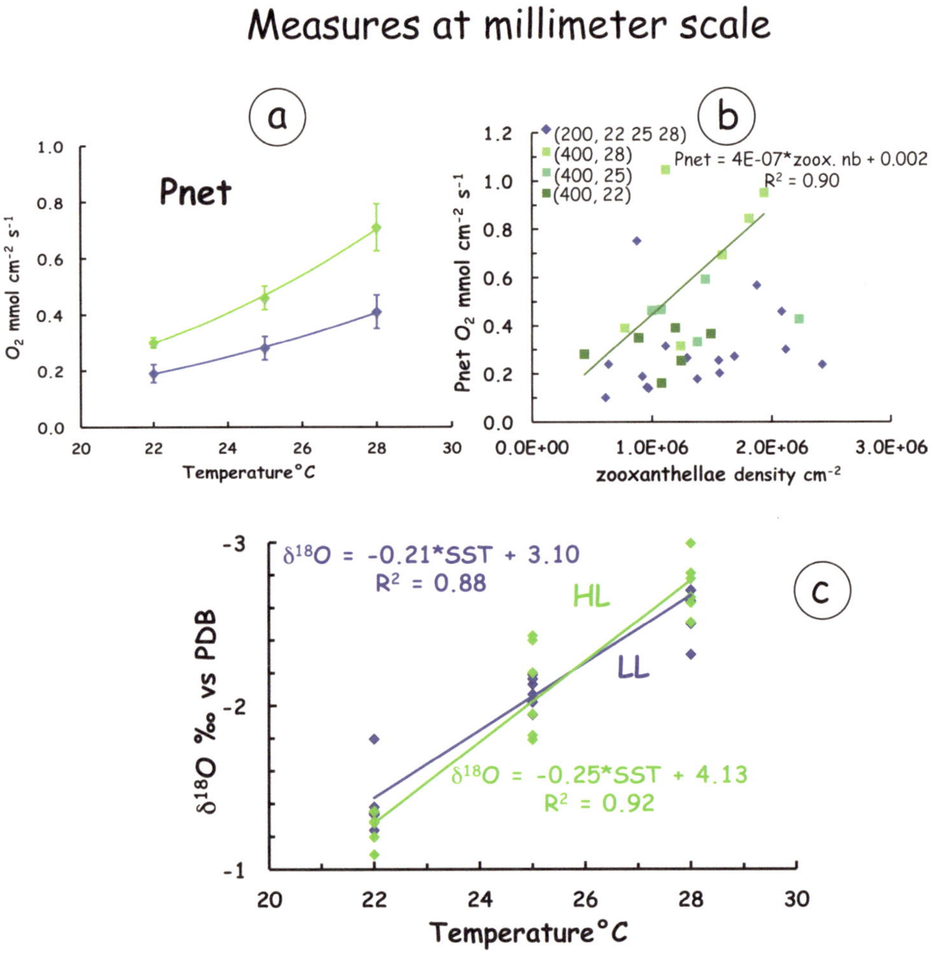


**Figure 2.**
*Test using a factorial design of three temperatures (22, 25, and 28°C) and two light intensities (200 and 400 µmol photon $m^{-2}$ $s^{-1}$) of cultured* Acropora. *Biologic response of net productivity (a), zooxanthellae density according to net productivity (b) and averaged $\delta^{18}O$-temperature calibration under high light (HL) and low light (LL) (c).*

## 2.2 Microstructures of coral skeletal

It was admitted that the coral skeleton such as coral *Acropora* (**Figure 3a** and **b**) or coral *Porites* presented composite mineral microstructures: centres of calcification (COC) and fibres, embedded in a few organic matter as a network [18]. COC were massive randomly oriented crystals called fusiform crystals (**Figure 3c, h** and **i**) [19], and numerous needle-like crystals projecting in many directions from the fusiform crystals were called the fibres gathered into bundles (**Figure 3e, f** and **g**) [19] oriented perpendicularly to the growth axis (**Figure 3e**) [19].

Skeletogenesis could be the result of two different processes: the deposition of fusiform crystals and the progressive thickening of the initial framework by needle like crystals (**Figure 3f**) [19].

These crystalline elements were differently distributed according to morphology [19–21]. Each microstructure is preferentially present in some morphological parts, which were more or less developed following the genus [22]. However, we know that they are composed by identical microstructures and only differ by their relative amounts.

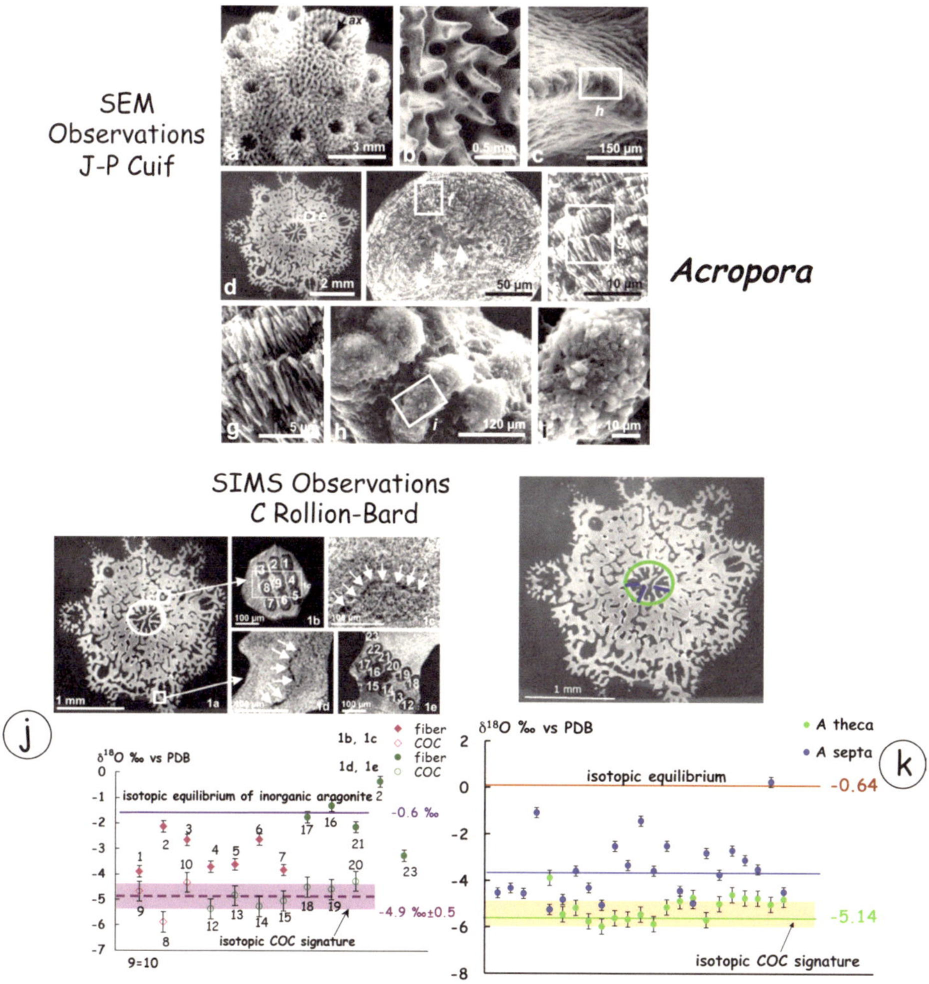


**Figure 3.**
*Identification of mineralisation mechanism of microstructures COCs and fibres using microscopic $\delta^{18}O$ analyses. SEM observations of cultured* Acropora *microstructures (a–i). microstructure identification and isotopic signatures (j). COC and fibre location on theca and septa from cultured* Acropora *(k).*

## 2.3 Consequences on $\delta^{18}O$

The oxygen isotopic composition of coral skeleton was measured on conventional spectrometer at mm size scale and might be also measured at micrometre scale by using ion microprobe.

Coral skeleton $\delta^{18}O$ was impacted by biology, which was essentially responsible of the vital effect. Most of the models proposed by geochemists neglect biology effects on $\delta^{18}O$, isotopic fractionation only depending on seasurface temperature and isotopic composition, $\delta^{18}O_{seawater}$ following Eq. (1). All climate reconstructions are derived according to this rule, including the estimate of both temperature and salinity based on the use of paired $\delta^{18}O$ and Sr/Ca measured on the same sample [23–25]. Salinity values deduced by this method are systematically misleading, d18O and Sr/Ca SST calibrations being based on classical thermodynamics. Whereas consequences of temperature and light on coral growth rates is well known by biologists [26], light effect is ignored by geochemists because the demonstration of this influence cannot be established from field data and needs culture tests.

$\delta^{18}O$ differed following microstructures at microscopic size scale (**Figure 3j** and **k**) [27]. This was confirmed later on [28, 29]. COC $\delta^{18}O$ was lower, while fibre $\delta^{18}O$ was higher and variable between equilibrium and COC value (**Figure 3j** and **k**) [29].

## 3. Calibrations of annual and monthly $\delta^{18}$O-temperature

### 3.1 Weber and Woodhead (1972) data set and annual calibrations revisited

Weber and Woodhead [3] established a formula able to predict past SST following the isotopic thermometer concept [30], expressed as:

$$SST°C = a \times \delta^{18}O\ (‰) + b \qquad (2)$$

(a) and (b) being constants (instead of A and B in [3]).

Because *Acropora* and *Porites* are ubiquitous genera, *Acropora* and *Porites* $\delta^{18}O$ calibration deriving from 835 and 421 sample, respectively, calibrations (**Figure 4a** and **b**) might be regarded as statistically significant. Moreover, isotopic analyses were conducted on annual samples, identified by X-ray growth bands, a pair of clear and dark bands corresponding to the annual growth [31]. Each temperature

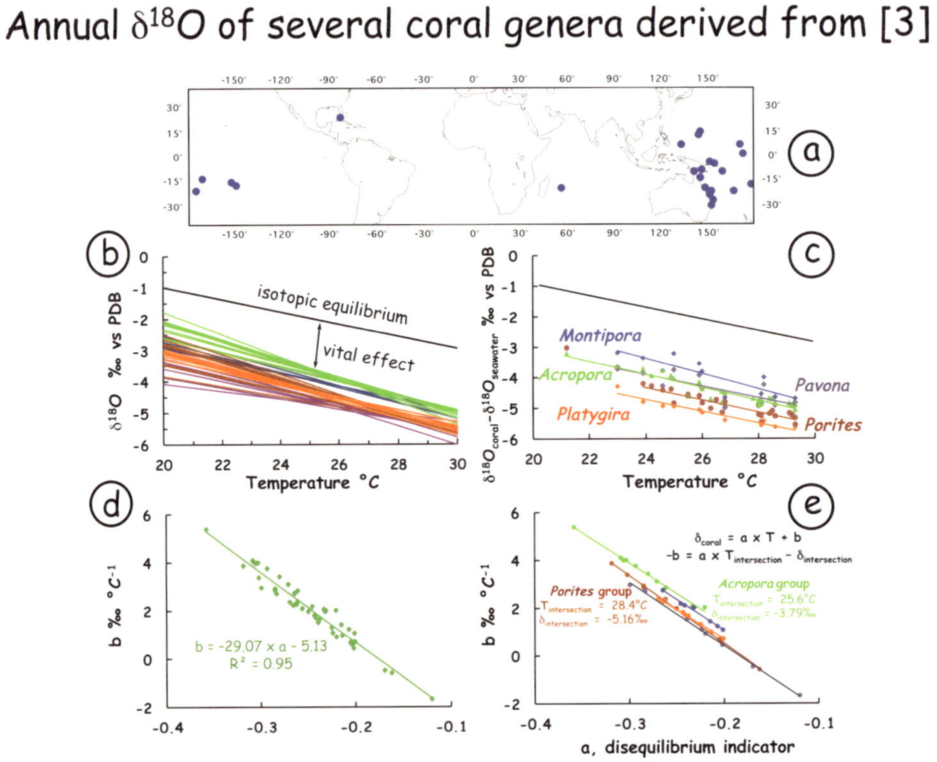


**Figure 4.**
*Annual coral $\delta^{18}O$ measures performed on several genera from revisited dataset of [3] (a–e). Location of the 29 sites where coral samples were collected, prescribing temperature values (a). Calibrations of 44 coral genera following $\delta^{18}O(‰) = a \times SST°C + B$ (b). The colours highlight genera sharing identical temperature and isotopic ranges, underlining the convergence of the groups including* Acropora *and* Porites. *Calibrations of some coral genera* Acropora, Porites, Platygira, Montipora *or* Pavona *following $\delta^{18}O_{carbonate} - \delta^{18}O_{seawater} = a \times \delta^{18}O(‰) + b$ (c). The colours of the calibrations correspond with colours from (c). strongly significant linear relationship linking constants a and b, of annual $\delta^{18}O$-annual temperature calibrations for 44 coral genera (d). Strongly significant linear relationship linking constant of annual $\delta^{18}O$-annual temperature calibrations for groups highlighted on (c and e).*

value, corresponding to one island, as associated to the averaged $\delta^{18}O$ measured for corals of the same species, all receiving identical local irradiation. Groups of genera including *Acropora* or *Porites* displayed strong convergence.

The dataset [3] did not take into account $\delta^{18}O_{seawater}$. Juillet-Leclerc and Schmidt [32] included annual $\delta^{18}O_{seawater}$ values assessed in the calibration established for *Porites* following the formula:

$$\delta^{18}O_{carbonate} - \delta^{18}O_{seawater} = -0.20 \times SST\,(°C) + 0.45 \quad (3)$$

with $R^2 = 0.83$, $N = 22$, $p < 0.001$, only significant over the SST range from 24 to 30°C [32].

However, the correlation linking $\delta^{18}O$ directly to temperature showed a higher coefficient [3]:

$$\delta^{18}O_{carbonate} = -0.27 \times SST\,(°C) + 2.24 \quad (4)$$

with $R^2 = 0.91$, $N = 24$, $p < 0.001$ including the lowest temperatures neglected in Eq. (5) [32].

A similar procedure was conducted for *Acropora*, using the same $\delta^{18}O_{seawater}$. We obtained:

$$\delta^{18}O_{carbonate} - \delta^{18}O_{seawater} = -0.21 \times SST\,(°C) + 1.26 \quad (5)$$

with $R^2 = 0.87$, $N = 24$, $p < 0.001$, significant over the temperature range from 21 to 30°C (**Figure 4e**).

the correlation linking $\delta^{18}O$ directly to temperature showed a higher coefficient [3]:

$$\delta^{18}O_{carbonate} = -0.28 \times SST\,(°C) + 3.34 \quad (6)$$

with $R^2 = 0.98$, $N = 27$, $p < 0.001$, significant over the temperature range from 21 to 30°C.

Slopes (a) shown by *Porites* and *Acropora* temperature calibrations including $\delta^{18}O_{seawater}$, −0.20 and −0.21‰/°C, respectively, differed from those deriving only from $\delta^{18}O_{carbonate}$ (referred as $\delta^{18}O$ in the following) and temperature. They were close to the slope of −0.19‰/°C assessed for inorganic aragonite calibration [33]. Slopes have been also obtained from other genera such as *Platygira*, *Montipora* or *Pavona* (**Figure 4e**) [3].

After introducing $\delta^{18}O_{seawater}$ into dataset [3], for *Porites* and *Acropora* genera, the usual thermodynamic equation is significant but to a lower degree, compared to Eqs. (4) and (6). For example, by taking into account only temperature, $R^2 = 0.91$ and 0.98 instead of 0.87 and 0.93 for *Porites* and *Acropora* respectively.

In the calibrations depending only on temperature, temperature might act first, according to thermodynamic law [5, 15] and second, through the photosynthetic process [34] (**Figure 2a**), which was enhanced by a temperature increase. Therefore, an increase in temperature induced a decrease in $\delta^{18}O$ following the first process while the second mechanism caused a rise in $\delta^{18}O$, confusing the global isotopic effect. Temperature influences $\delta^{18}O$ twice, explaining that temperature is the main factor, which does not exclude the role of $\delta^{18}O_{seawater}$.

Calibrations taking into account $\delta^{18}O_{seawater}$ exhibited a slope value close to isotopic equilibrium of inorganic aragonite with water, suggesting that under quasi-uniform light, the isotopic offset of coral $\delta^{18}O$ is constant, regardless of temperature (**Figure 4c**). Eqs. (3) and (4) confirm that, to a lesser degree than temperature, $\delta^{18}O_{seawater}$ may be included in a calibration.

When comparing constants (a) and (b) of Eq. (2) from data series [3], for all genera annual $\delta^{18}O$-annual temperature calibrations (**Figure 4d**), we obtained a strongly significant linear relationship:

$$b = -29.07 \times a - 5.13 \quad (7)$$

with $R^2 = 0.95$, $N = 37$ and $p < 0.001$. (a) corresponds to a disequilibrium indicator compared to −0.19, the slope value derived from the theoretical $\delta^{18}O$ at equilibrium [34]. Such a relationship was not hazardous, but reflected inherent features of annual $\delta^{18}O$-annual SST calibrations. Linear calibrations determined from single genus deduced from figures or table of [3], showed strong correlation coefficients: $R^2 = 0.99$ (**Figure 4e**).

This suggests that the $\delta^{18}O$ SST dependence is based on a unique rationale according to taxonomy, in turn inherent to the coral skeleton.

Dataset [3] revealed unique relationship between annual $\delta^{18}O$-annual temperature calibrations of each genus, because coral taxonomy is based on morphology. Land et al. [22] stressed the high $\delta^{18}O$ variability following the longitudinal section on the calices of some species or the septa dentations of another one, inducing that according to coral morphology, some skeleton portions might be more or less developed, implying a large isotopic variability.

We underlined the relationship existing between the annual $\delta^{18}O$-annual SST calibration constants. However, identical feature was highlighted for annual Sr/Ca-annual temperature calibrations [35–37]. The link existing between $\delta^{18}O$ and Sr/Ca is not straightforward, oxygen being a component of $CaCO_3$ and Sr/Ca an impurity included in the skeleton. However, it is possible to recognise common $\delta^{18}O$ and Sr/Ca behaviour relative to their microstructure distribution in the coral and the concept of taxonomy.

Coral skeleton presents composite mineral microstructures: centres of calcification (COC) and fibres, embedded in a few organic matter as a network [18], differently distributed according to morphology [19–21]. Latter authors show that COC and fibres are essentially present in morphological parts, more or less developed following the genus. On the one hand, $\delta^{18}O$ signature differs according to the microstructures [19–21], COC $\delta^{18}O$ being lower than fibre $\delta^{18}O$ [29] (**Figure 3j** and **k**). On the other hand, Sr/Ca ratios measured on COCs are higher than those of fibres [9]. Cohen et al. [37] examined synchronously deposited microstructures on *Porites lutea* over a year, exhibiting COC elemental ratios systematically higher compared to those of fibres developed over an identical period. Thus, it is the proof that annual COC Sr/Ca value is higher than the annual fibre Sr/Ca signature [37]. Therefore, we suggested that discrepancies of morphology existing between coral genera are due to differences of microstructure proportions [22], explaining common features between the annual trace element ratio- and annual $\delta^{18}O$-annual temperature calibrations.

### 3.2 Identification of microstructures and their isotopic signatures using microsensor

Several small colonies of *Acropora verweyi* (Archaeocoeniina) were cultured following the procedure described by Reynaud-Vaganay et al. [38] under constant and controlled conditions, in Centre Scientific of Monaco (CSM) (**Figure 5**). Such colonies grew glued onto glass slides (**Figure 5**). Morphology of the microstructures, using on a scanning electron microscope (SEM Philips 505) was similar to observations performed by Cuif and Dauphin [19] (**Figure 3a–i**).

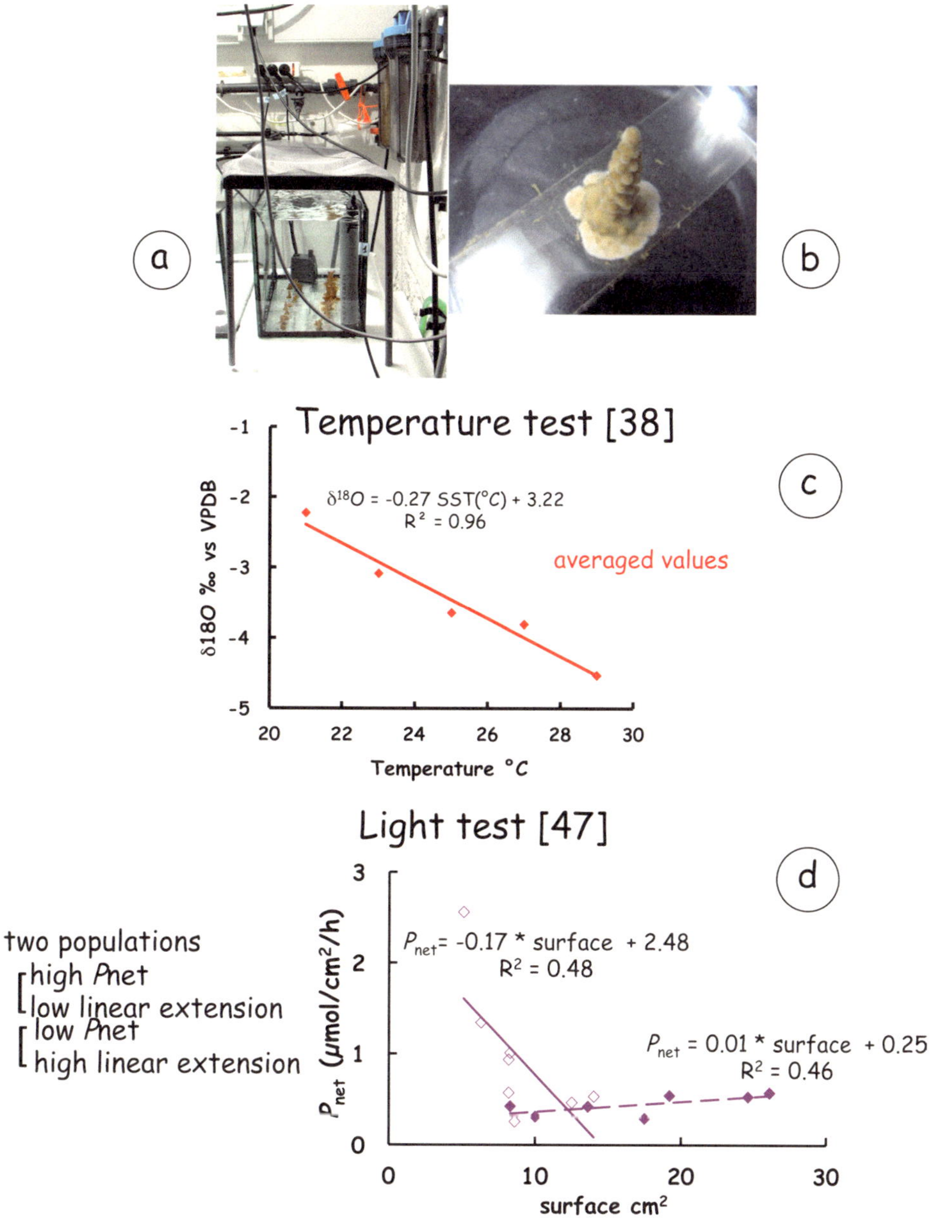


**Figure 5.**
*Culture experiment (a and b) to test $\delta^{18}O$ variability caused separately by temperature (c) and by light (d). Aquarium in CSM (a) coral* Acropora *glued on glass slide showing new formed aragonite both on the colony surface and on the glass slide (b). $\delta^{18}O$-temperature calibration derived from averaged isotopic measures (c). Relationship between Pnet and linear extension, revealing the partition of colonies into two populations, following potential photosynthetic response of colonies (d).*

The new skeleton, formed under unique controlled condition, was grown on the glass slide, and sampled for the calibration of the growth units (**Figure 3j**). COC- and fibre-enriched zones were identified using SEM [29]. To characterise separately the isotopic signature of fibres and COC, analyses were focused on the microstructures earlier identified on the newly formed skeleton on two zones (**Figure 3a–c**; **Figure 3j**). We then focused our measurements around the theca of the newly formed skeleton (**Figure 3k**) where Gladfelter [39] recognised large amounts of "fusiform crystals" (**Figure 3e**). The sampling of the septae aimed at confirming the presence of both COC and fibres as they have been identified from SEM observations [26].

The present study, confirmed that there was a strong relationship between isotopic value, crystal shape and skeleton morphology [29]. Crystals called "fusiform" by Gladfelter [39], according to their shape, show the same isotopic values as COC. We distinguished in septa both COC and fibres (**Figure 3k**). This confirmed microscopic observations of septa [21] showing discontinuous COC surrounded by fibres.

Isotopic fractionation was likely of kinetic origin, the rate changing according to microstructure. Skeleton microstructures $\delta^{18}O$ shed in light how chemical and/or physical processes might be adapted by biology to form crystals characterised by specific shapes and distributed following a hierarchical arrangement. The present study demonstrated that the presence of organic molecules (the organic matrix located at the interface tissue-mineral (**Figure 1**) had the capability to control the mineral deposition mechanism. Probably, the influence of external factors should be superimposed on the chemical signature of coral biomineral genetically determined).

### 3.3 Monthly calibrations on coral *Porites*

The preliminary step of climatic reconstruction using *Porites* skeleton, the genus more often analysed in this context, consisted of the assessment of seasonal $\delta^{18}O$-seasonal temperature calibration based on monthly instrumental temperatures over the last decades covered by the core. Sampling was conducted along the coral's growth through time, following the maximal growth rate perpendicular to the annual density bands shown by X-ray [40].

At millimetre size scale, it was also possible to highlight the strongly significant linear relationship between the constants of seasonal $\delta^{18}O$-seasonal temperature calibrations and to relate behaviour of the constants of the seasonal-$\delta^{18}O$- and Sr/Ca-seasonal temperature calibrations to the presence of two crystallographic units. Following DeLong et al. [40], fibres insuring the thickening of a colony should be preferentially deposited during a less active photosynthesis, whereas COC insuring axial growth should be formed during high photosynthetic activity. Juillet-Leclerc and Reynaud agreed, however, they demonstrated that growth mode was not so simple [34].

In order to test seasonal $\delta^{18}O$-seasonal temperature calibration variability including the seasonal light effect, calculated for several coral cores collected on a given site, at different temperature ranges, we considered studies conducted on several *Porites* colonies from three sites. The mean annual temperature offshore Amédée Island, New Caledonia (22°29′ S, 166°28′ E) was 24.72°C, over the period 1968–1992 [41, 42], while at Clipperton Atoll (10°18′ N, 109°13′ W) the mean annual temperature was 28.5°C, over the period 1985–1995 [43] and in the Flores Sea, Indonesia (6°32′ S, 121°13′ E) the mean annual temperature was above 28°C, over the period 1979–1985 [44].

We assumed that calibrations measured on different coral colonies grown at a given site (New Caledonia, Clipperton or Indonesia) differed according to various light sensitivities due to depth or light incidence or acclimation because seasonality strongly affected light variations, and was likely different following site location (**Figure 6a**). However, calibration constants calculated from monthly data for *Porites* remained strongly correlated (**Figure 6b**) as we previously observed for annual $\delta^{18}O$-annual temperature calibrations.

As seasonal $\delta^{18}O$-seasonal temperature calibrations presented similar behaviours, even in different sites characterised by distinct $\delta^{18}O_{seawater}$, they did not reflect classical thermodynamics.

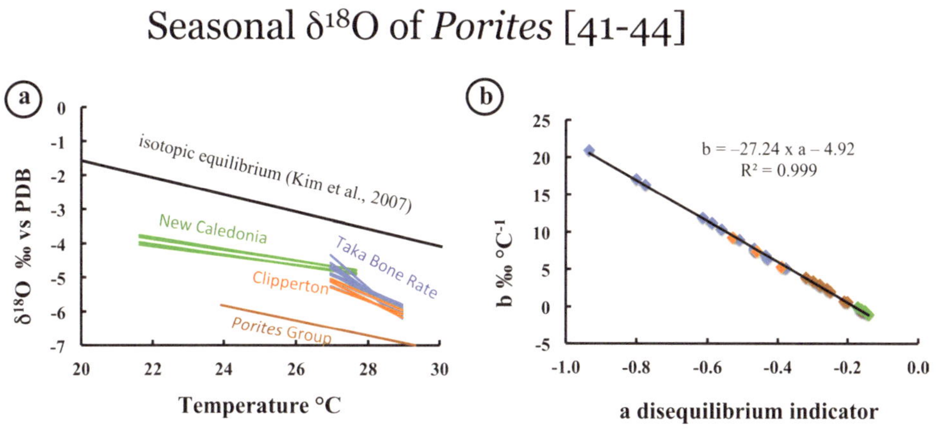


**Figure 6.**
Porites *monthly $\delta^{18}O$ from [41–44] (a and b). Monthly $\delta^{18}O$-monthly temperature calibrations for coral* Porites *and annual $\delta^{18}O$-annual temperature for* Porites *group as defined in* ***Figure 4c*** *(a) and associated constant relationship (b).*

## 4. Coral cultures simulating different environmental conditions

### 4.1 Test of temperature

New technique of culture was developed to calculate $\delta^{18}O$-temperature calibration for *Acropora sp*. The experiment was conducted in Centre Scientific of Monaco (CSM) using colonies of the branching zooxanthellate scleractinian coral, *Acropora sp*. (**Figure 5a** and **b**), in the Gulf of Aqaba (1 m depth) [38]. The nubbins (new colony fragments) were collected from unique parent colony. The specimens were glued on glass slides. Chemical conditions were kept constant during the experiment, as $\delta^{18}O_{seawater}$ (1.29 ± 0.01‰ vs. SMOW) measured in the aquaria under light (260 μmol $m^{-2}$ $s^{-1}$) on a 12:12 h photoperiod [38]. Five temperatures were tested on six coral fragments. The skeletal powder was treated following the method described by Boiseau and Juillet-Leclerc [45].

The calibration given by the experiment (**Figure 5c**) might be expressed as:

$$\delta^{18}O_{coral} - \delta^{18}O_{seawater} = 0.27 \times SST(°C) + 3.22 \quad (8)$$

with N = 5 and $R^2$ = 0.96.

As $\delta^{18}O_{seawater}$ = 1.29 vs. SMOW = 1.02 vs. PDB was constant, the calibration might be expressed as:

$$\delta^{18}O_{coral} = -0.27 \times SST(°C) + 5.41 \quad (9)$$

### 4.2 Test of light intensity

Heterotrophy and photosynthesis were linked and were difficult to separate in field experiments. Coral cultures enabled the investigation of each parameter at a time [34, 46].

Tips from 24 branches were sampled from a single parent colony of *Acropora sp*. All colonies were cultured for 6 weeks under a light intensity of 130 μmol $m^{-2}$ $s^{-1}$. The ring skeleton deposited on the glass slide was then removed with a scalpel [38] dried overnight at room temperature and stored in glass containers pending isotopic analyses. Thereafter, colonies were cultured for an additional period of 6 weeks under a light intensity of 260 μmol $m^{-2}$ $s^{-1}$, and their isotopic composition was determined (**Figure 5d**). The extension of new aragonite on the glass slide was assimilated to linear extension.

In the first time [46], the averaged results showed that daily calcification, net photosynthesis significantly increased with increasing light and skeletal $\delta^{18}O_{coral}$ were more negative under low light than high light, −4.2 versus −3.8.

Another interpretation, considering each colony was later published [34]. Following the evolution, $\delta^{13}C_{coral}$ increasing or decreasing, two populations appeared: one responding to light with increasing net photosynthesis associated to low linear extension and the other characterised by poor net photosynthesis associated to high linear extension (**Figure 5d**).

We suggested that different behaviours were due to different zooxanthellae amounts contained by colonies.

### 4.3 Factorial design of three temperatures and two light intensities

Conditions applied to each tank were referred in the following as (light in μmol photons $m^{-2}$ $s^{-1}$, temperature in °C): (200, 22), (200, 25), (200, 28), (400, 22), (400, 25) and (400, 28) [15, 48].

Culture procedures are similar to what was described in **Figure 5a** and **b**. Responses of photosynthesis and zooxanthellae density are displayed in **Figure 2**.

Calibrations calculated from the mean $\delta^{18}O$ values for each temperature regime were consistent with those previously published (**Figure 2c**) [49, 50]. $\delta^{18}O$ *versus* temperature calibrations of nubbins cultured under LL and HL were both highly significant ($R^2$ = 0.94, N = 18, P = 0.001 and R = 0.96, N = 18, P = 0.001, respectively). The slope value was in good agreement with Eq. (2) commonly used for *Porites* corals with b = −0.20‰/°C [37, 51, 52]. However, the values obtained at 22, 25 and 28°C showed a large scattering both at LL and HL, from 0.5 to 1.25‰ or the equivalent of 2 to 5°C.

By comparing our results with other culture experiments [51–53], differences appeared between various $\delta^{18}O$-temperature (°C) calibrations regarding both the slopes and the intercepts with the temperature scale. We suggested that this could be due to inter-species or inter-colony $\delta^{18}O$ differences. Even two calibrations obtained on cultured *Acropora* exhibited differences (the present study and that of Reynaud et al.) [52].

Mean $\delta^{18}O$ values calculated for each temperature did not vary with light, which contradicted observations made of mean physiological parameters, in contrast with other proxies ($\delta^{13}C$, Sr/Ca and Mg/Ca). By changing the light intensity from low to moderate, Juillet-Leclerc and Reynaud [47] recorded a $\delta^{18}O$ increase associated with skeletal infilling following a kinetic process. The present experiment, conducted under high light intensities, did not show a similar behaviour. We suggested that temperature and light effects on isotopic composition were competing. The results of the present experiment indicated that, under the chosen conditions, the temperature effect was more important than the light effect. This was illustrated by the weak discrepancy in the mean $\delta^{18}O$ recorded at 28°C.

Previous culture experiments have been conducted to test the temperature effect on $\delta^{18}O$ [38, 51–54]. Due to the sensitivity of the photosynthesis to temperature, $\delta^{18}O$-temperature calibrations will always include temperature-dependent photosynthetic changes, enabling the vital effect due to temperature only to be deconvolved from the total signature. Therefore, all calibrations, even established on a single coral head or from cultured nubbins, are impacted by photosynthetic activity linked to zooxanthellae density. The universal calibration does not exist.

## 5. A new paradigm for $\delta^{18}O$ in coral skeleton oxygen isotope fractionation response to biological kinetic effects

During the last experiment, we failed understanding the light effect on $\delta^{18}O$. We kept one colony cultured in each of six light and temperature conditions previously discussed. Knowing that standard error obtained in the first step of the experiment for six samples cultured in the same condition was between 0.02 and 0.12, we consider that values measured on one colony by using SIMS were representative for each environmental condition [48].

We discussed our results after listing all biological and biological advances, such as (i) conclusions derived from inorganic $CaCO_3$ precipitation disturbed by biology, biased by non-realistic models [8, 55, 56]; (ii) the potential role of the calicoblastic layer composed of proteins, sugars and water [57, 58] and (iii) the role of carbonic anhydrase (CA), ubiquitous enzymes known to act as catalysts for the interconversion of $CO_2$ and $HCO_3^-$ [59, 60].

### 5.1 Material

The random SIMS measurements were made exclusively on the newly formed skeleton coenosteum (the skeleton portion separating corallites), avoiding newly formed corallites and spines [21]. Samples were distributed along the growth axis from the initial branch to the rim of the expanded tissue (**Figure 7a**).

Measurements were performed using the Cameca IMS 1280-HR ion microprobe at the CRPG, and the comparison of SIMS and conventional mass spectrometer measurements were made using data from [29, 48].

The $\delta^{18}O$ SIMS measurements are displayed as histograms, with bin width of histograms, 0.25‰, depending on the precision of the measurements between 0.09 and 0.32‰ (1σ) (**Figure 7b**).

### 5.2 Isotopic results

At 22°C, although the mean values were identical within the analytical error, light had a significant effect on the $\delta^{18}O$ distribution. Under LL, the single high bar was surrounded by values spread over 2‰ (**Figure 7b**). By contrast, under HL, values spread over 4‰, $\delta^{18}O$ distribution was bimodal, two high bars were observed, one bar centred on −0.01‰, followed by decreasing values that exceeded the expected $\delta^{18}O$ values for aragonite precipitated in oxygen isotope equilibrium in water.

At 25°C, the bimodal $\delta^{18}O$ distribution in the two samples exhibited two high bars, with the more depleted in $^{18}O$ peak being the same in the two light conditions (**Figure 7b**).

At 28°C, $\delta^{18}O$ distribution was bimodal, the isotopic amplitude being slightly higher under LL than under HL (**Figure 7b**).

Assuming that two high bars observed in histograms were significant, we used Ashman's D test [61] to strengthen the bimodality of the LL-25°C, LL-28°C, HL-25°C and HL-28°C.

### 5.3 Discussion

The histograms showed bimodal distributions (except for the colony grown in LL-22°C) caused by distinct kinetic processes. All the colonies were submitted to the diurnal cycle of 12 h light and 12 h dark necessary to grow healthy coral. Therefore, we assumed that, for the high isotopic bar corresponding to the values more depleted in $^{18}O$ (**Figure 2g**), corresponding to the highest kinetic

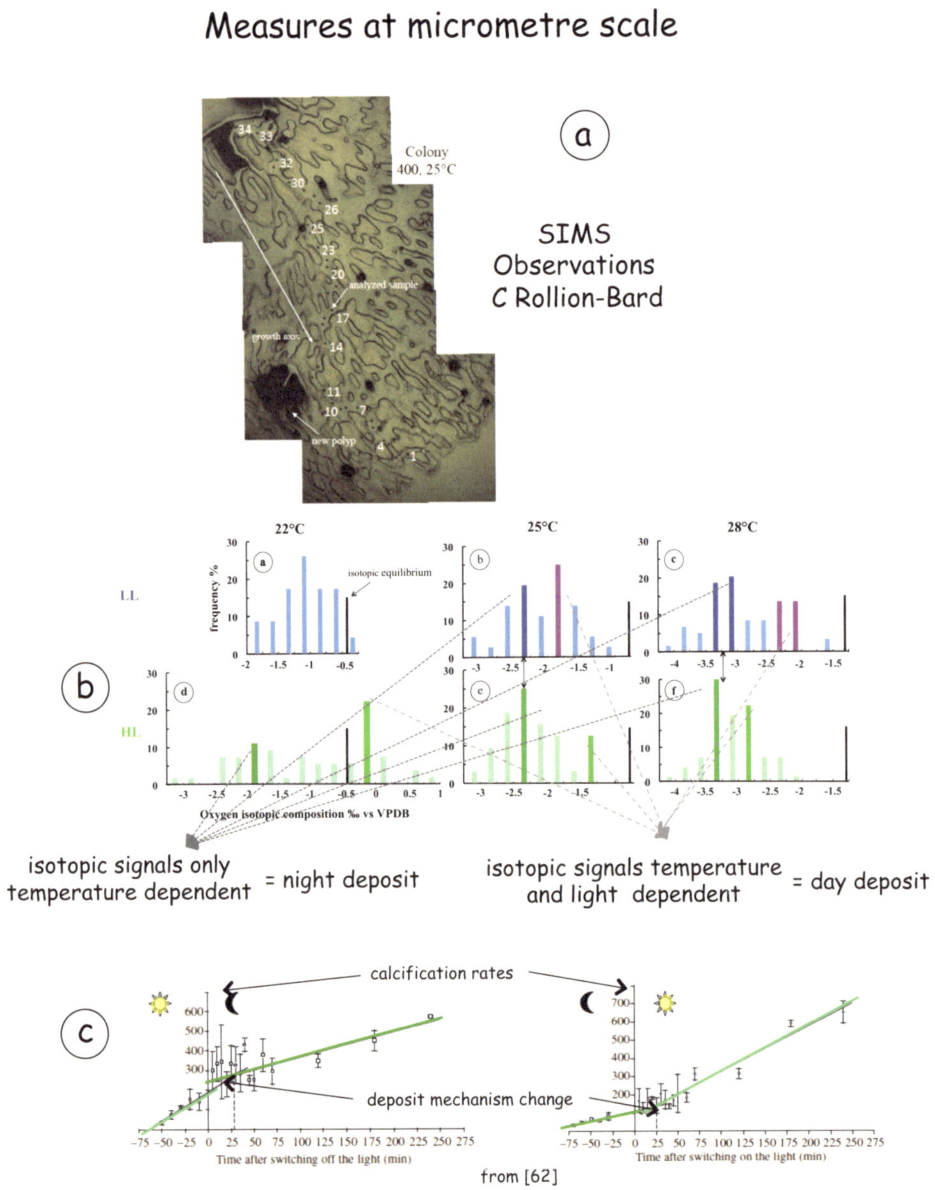


**Figure 7.**
*Test using a factorial design of three temperatures (22, 25, and 28°C) and two light intensities (200 and 400 µmol photon $m^{-2}$ $s^{-1}$) of cultured* Acropora*. SIMS observations of sampling of microscopic scale analyses (a). Isotopic responses displayed as histograms for each environmental conditions showing bimodal distribution (b) corresponding to nighttime and daytime calcification [58] (c).*

fractionation, only depending on temperature is associated with nighttime. By contrast, for the other high isotopic bar corresponding to the values less depleted in $^{18}O$, in turn to the weakest kinetic process, depending both on temperature and light could be associated with daytime calcification (**Figure 7b** and **c**).

By culturing *Stylophora pistillata* in controlled conditions similar to those in our experiment [62] with a diurnal cycle of 12 h light and 12 h dark, the authors measured calcification and observed that the calcification rate differed according to night and day conditions (**Figure 7b** and **c**). The regressions showed that the light calcification rate was about 2.4 times higher than the dark calcification rate (**Figure 7c**). However, when conditions shifted from light to dark or from dark to light, the calcification rate experienced a lag of 25 min between the dark or light regression relative to time (**Figure 7c**). The lag was likely due to the change of calcification process, which differed between nighttime and daytime. These biological evidences were in good agreement with our biochemical ones (**Figure 7b** and **c**). Two assumptions are proposed to explain such a mechanism: a pH change in the extracellular calcifying medium (ECM) and a modification of the biochemical compounds of the organic matrix [62].

Therefore, assuming that the dual high $\delta^{18}O$ bars exhibited in the histograms (**Figure 7b**) were related to the $Ca^{2+}$-pump activity, this could modify the internal pH [62, 63]. If the pH is lower in the dark than in the light [62, 63], in line with McCrea's [65] calculations, illustrated by Adkins et al. [52], high $\delta^{18}O$ bars less depleted in $^{18}O$ should be associated with nighttime calcification, and high $\delta^{18}O$ bars more depleted in $^{18}O$ should correspond to daytime conditions (**Figure 7b** and **c**).

However, our experiment demonstrated an opposite distribution: High $\delta^{18}O$ bars less depleted in $^{18}O$ were identified as daytime skeleton deposits, and those more depleted in $^{18}O$ were identified as nighttime deposits. Therefore, we concluded that the two distinct high $\delta^{18}O$ bars could not be caused by internal pH diurnal variations. In contrary to the common assumption [9], we demonstrated that mineralisation is not controlled by classical thermodynamic rules, that is, pH, but rather should obey biological kinetic effects, following a mechanism that remains to be identified.

Now we need to identify a mechanism that allowed daytime and nighttime mineralisation to be distinguished, knowing that the second assumption given by Moya et al. [62], a modification of the biochemical compounds of the organic matrix, remained to be examined.

### 5.4 Influence of organic matter on crystallisation

It was supposed that the photosynthetic supply of precursors might modify the biochemical composition of an organic matrix [62], which necessitated an internal rearrangement related to the secretion of specific proteins defending the observed lag. The formation of the organic matrix, controlled by the calicoblastic cells, appeared to be a prerequisite for crystallisation [58–67] Recently, 36 proteins were extracted from the skeletal organic matrix (SOM) embedded within aragonite crystals, constituting a bio mineralisation toolkit including at least two Carbonate Anhydrase [68] accelerating mineralisation. From this toolkit, four unique proteins, coral acid-rich proteins (CARP), catalysed the precipitation of $CaCO_3$ in vitro [68]. Moreover, some proteins appeared to be differentially expressed between day and night [68, 69]. Therefore, the two different proteins caused different kinetic fractionation processes, inducing during the night higher kinetic isotope fractionation than during the day. We note that calcification rate and isotope fractionation kinetics were drastically different concepts.

Results derived from our last geochemical experiment should drove to responses also addressed by biological study. The fact that classical geochemistry rule, such as pH, could not explain isotopic behaviour led to look for another assumption. Therefore, it highlighted that coral mineralisation could be controlled by proteins secreted by organic matrix. This evidence is now well admitted, supported by multiple biological studies [68, 69].

## 6. Simple models

### 6.1 Model according to microstructure distribution

As early as 1982, Gladfelter [39] assumed that linear extension and infilling were two independent growth rates, an assumption supported by Juillet-Leclerc and

Reynaud [47]. The authors demonstrated that each growth rate was related to preferential deposition of microstructures, COCs ensuring linear extension and fibres, infilling.

Furthermore, geochemical investigations revealed that crystal isotopic signatures differed [27–29, 48]. COC formation should be related to temperature [39] and fibre deposit depends on both temperature and light [48]. Therefore, temperature and light changes interplayed to determine skeletal isotopic composition.

Sampling conducted as it was described by DeLong et al. [40] included both COCs and fibres. Changes of relative amounts of microstructure as illustrated by X-rays and their respective $\delta^{18}O$ were determined by their mechanisms of formation, unknown so far [29]. Following isotopic laws, the combination of calcification processes and isotopic fractionation could be expressed as:

$$\text{measured } \delta^{18}O = \left[\left(x_{COC} \times \delta^{18}O_{COC}\right) + \left(x_{fibre} \times \delta^{18}O_{fibre}\right)\right] / \left(x_{COC} + x_{fibre}\right) \quad (10)$$

where $x_{COC}$ and $x_{fibre}$ are the relative amounts of the crystal microstructures, with $x_{COC} + x_{fibre} = 1$, and $\delta^{18}O_{COC}$ and $\delta^{18}O_{fibre}$ are their isotopic signatures depending on temperature and temperature and light, respectively. This expression is likely to be simplistic but closer to the truth than the thermodynamic formula. Temperature is the prominent factor because included both in the crystal amounts and the isotopic signatures. $SST_{intersection}$, the corresponding $\delta^{18}O_{intersection}$, should be related to morphology [22]. When using Eq. (10), the intersection of calibration should be obtained when $\delta^{18}O_{intersection} = (0.50 \times \delta^{18}O_{COC}) + (0.50 \times \delta^{18}O_{fibre})$ or at $SST_{intersection}$, $\delta^{18}O_{intersection} = (\delta^{18}O_{COC} + \delta^{18}O_{fibre})/2$. As long as temperature does not reach $SST_{intersection}$, more fibres are formed in the coral skeleton and when temperature exceeds $SST_{intersection}$, COC are progressively prevailing.

## 6.2 Model according to environmental parameters

In Pacific Ocean, local zones may be characterised by seasonal and/or interannual environmental parameter amplitude, as ΔSST. By this way, we are able to identify El Niño-Southern Oscillation (ENSO) occurrence [70], over past time.

$\Delta SST < 2°C$, seasonal conditions occurring in Tarawa atoll, in Galapagos or in Fiji [71], $\delta^{18}O$ seasonal variability mimics SSS variability. In Fiji, $\delta^{18}O$ is correlated to seasonal precipitation [72]. In other sites, $\delta^{18}O$ variability may indicate oceanic advection. Such events are directly related to El Niño.

If seasonal $\delta^{18}O$ is recorded over several decades, interannual variability may be isolated. By removing the seasonal cycle and applying a 13-month running mean filter from monthly $\delta^{18}O$, interannual isotopic variability may be regarded as temperature, the greatest fluctuations revealing El Niño-Southern Oscillation (ENSO) occurrence [70], or the global warming over the twentieth and twenty-first century.

When $\Delta SST \geq 5°C$, as it is occurring off South Korea coast [72], temperature and $\delta^{18}O_{seawater}$ are both involved in coral skeleton $\delta^{18}O$ variability. Environmental parameters are difficult to separate. If $\delta^{18}O$ shows strong decrease associated to great SST drops, it may be caused by the occurrence of La Niña (characterised by colder SST than the normal conditions) [70].

When $2°C \leq \Delta SST \leq 4°C$ as it is recorded in the central tropical Pacific as in Palmyra [73], $\delta^{18}O$ snapshots focused on crucial past periods demonstrate that a 2- to 7-yr bandpassed record (the lower-frequency of ENSO) [70], the interannual isotopic signal highlights El Niño and La Niña occurrences.

Coral skeletal $\delta^{18}O$ is well-suited tool to shed in light climatic events before and after industrial era, to predict future events in the next decades [70].

## 7. Conclusion

Coral skeleton $\delta^{18}O$ does not obey to classical thermodynamics but rather reflects aragonite microstructure distribution. We demonstrated that oxygen isotopic fractionation is essentially temperature dependent, due to two temperature effects, one following thermodynamic law, decreasing $\delta^{18}O$ when temperature increases, and second, temperature acting through photosynthetic process, increasing $\delta^{18}O$ when temperature increases. Consequently, when temperature changes, $\delta^{18}O$ is affected in opposite senses, confusing the global isotopic effect.

$\delta^{18}O$ measured at millimetre size scale on coral colonies cultured in controlled conditions under varying temperatures and/or light intensities, allows highlighting biologic and isotopic changes associated to environmental factors, acting as vital effect. Measured at microscopic size, $\delta^{18}O$ reveals mineralisation processes. By using the last method coupled with biologic evidences, the role of proteins and enzymes, secreted by organic matrix at the interface tissue mineral is demonstrated, showing the potential biologic control on meralisation.

## Author details

Anne Juillet-Leclerc
LSCE, Gif sur Yvette, France

*Address all correspondence to: anne.juillet-leclerc@lsce.ipsl.fr

## References


[1] Allemand D, Ferrier-Pagès C, Furla P, Houlbrèque F, Puverel S, Reynaud S, et al. Biomineralisation in reef-building corals: From molecular mechanisms to environmental control. General Palaeontology (Palaeobiochemistry) Comptes Rendus Palevol. 2004;**3**:453-467

[2] Salvat B. Le corail et les récifs coralliens. Ouest France: Découverte Nature Ed; 2003

[3] Weber JN, Woodhead PMJ. Temperature dependence of oxygen-18 concentration in reef coral carbonates. Journal of Geophysical Research. 1972;**77**:463-473

[4] Epstein S, Buchsbaum R, Lowenstam H, Urey HC. Carbonate water isotopic temperature scale. Bulletin Geological Society of America. 1951;**62**:417

[5] EpsteinS, BuchsbaumR, LowenstamH, Urey HC. Revised carbonate-water isotopic temperature scale. Bulletin Geological Society of America. 1953;**62**:417-425

[6] Weber JN, Woodhead PMJ. Ecological studies of the coral predator *Acanthaster planci* in the South Pacific. Marine Biology. 1970;**6**:12-17

[7] McConnaughey TA. C-13 and O-18 isotopic desequilibrium in biological carbonates: I. Patterns. Geochimica et Cosmochimica Acta. 1989;**53**:151-162

[8] McConnauguey TA. C-13 and O-18 isotopic disequilibria in biological carbonates: II. In vitro simulation of kinetic isotope effects. Geochimica et Cosmochimica Acta. 1989;**53**:163-171

[9] Cohen A, McConnaughey TA. Geochemical perspectives on coral mineralization reviews in mineralogy and geochemistry. In: Dove PM, de Yoreo JJ, Weiner S, editors. Biomineralization. Vol. 54. Mineralogical Society of America; 2003. pp. 151-187

[10] Gaetani GA, Cohen AL. Element partitioning during precipitation of aragonite from seawater: A framework for understanding paleoproxies. Geochimica et Cosmochimica Acta. 2006;**70**:4617-4634

[11] Cohen AL, Gaetani GA, Lundälv T, Corliss BH, George RY. Compositional variability in the cold-water scleractinian, *Lophelia pertusa*: New insights into "vital" effects. Geochemistry, Geophysics, Geosystems. 2006;7. DOI: 10.1029/2006GC001354

[12] Gagnon AC, Adkins JF, Fernandez DP, Robinson LF. Sr/Ca and Mg/Ca vital effects correlated with skeletal architecture in a scleractinian deep-sea coral and the role of Rayleigh fractionation. Earth and Planetary Science Letters. 2007;**261**:280-295

[13] Gaetani GA, Cohen AL, Wang Z, Crusius J. Rayleigh-based, multi-element coral thermometry: A biomineralization approach to developing climate proxies. Geochimica et Cosmochimica Acta. 2011;**75**:1920-1932. DOI: 10.1016/j.gca.2011.01.010

[14] Veron JEN. Corals of Australia and the Indo-Pacific. University of Hawaii Press; 1986

[15] Juillet-LeclercA, ReynaudS, DissardD, Tisserand G, Ferrier-Pagès C. Light is an active contributor to vital effects of coral skeleton proxies. Geochimica et Cosmochimica Acta. 2014;**140**:671-690

[16] Barnes DJ, Chalker BE. Calcification and photosynthesis in reef-building corals and algae. In: Dubinsky Z, editor. Coral Reefs. Ecosystems of the World. Vol. 25. Amsterdam: Elsevier; 1990. pp. 109-131

[17] Iluz D, Dubinski D. Coral photobiology: New light on old views. Zoologica. 2015;**118**:71-78

[18] Von Euw S, Zhang Q, Manichev V, Murali N, Gross J, Feldman LC, et al. Biological control of aragonite formation in stony corals. Science. 2017;**356**:933-938

[19] Cuif JP, Dauphin Y. Microstructural and physico-chemical characterisation of centres of calcification in septa of some scleractinian corals. Palaeontologische Zeitschrift. 1998;**72**:257-270

[20] Stolarski J. Three-dimensional micro- and nanostructural characteristics of the scleractinian coral skeleton: A biocalcification proxy. Acta Palaeontologica Polonica. 2003;**48**:497-530

[21] Nothdurft LD, Webb GE. Microstructure of common reef-building coral genera *Acropora*, *Pocillopora*, *Goniastrea* and *Porites*: Constraints on spatial resolution in geochemical sampling. Facies. 2007;**53**:1-26

[22] Land LS, Lang JC, Barnes DJ. Extension rate: A primary control on the isotopic composition of west Indian (Jamaican) scleractinian reef coral skeletons. Marine Biology. 1975;**3**:221-233

[23] Zinke J, Rountrey A, Feng M, Xie SP, Dissard D, Rankenburg K, et al. Corals record long-term Leeuwin current variability including Ningaloo Niño/Niña since 1795. Nature Communications. 2014. DOI: 10.1038/ncomms4607

[24] Tierney JE, Abram NJ, Anchukaitis KJ, Evans MN, Giry C, Kilbourne KH, et al. Tropical Sea surface temperatures for the past four centuries reconstructed from coral archives. Paleoceanography. 2015. DOI: 10.1002/2014PA002717

[25] Wu HC, Linsley BK, Dassié EP, Schiraldi B Jr, de Menocal BP. Oceanographic variability in the South Pacific convergence zone region over the last 210 years from multi-site coral Sr/Ca records. Geochemistry, Geophysics, Geosystems. 2013. DOI: 10.1029/2012GC004293

[26] Gattuso JP, Allemand D, Frankignoulle M. Photosynthesis and calcification at cellular, organismal and community levels in coral reefs: A review on interactions and control by carbonate chemistry. American Zoologist. 1999;**39**:160-183

[27] Rollion-Bard C, Chaussidon M, France-Lanord C. pH control on oxygen isotopic composition of symbiotic corals. Earth Plan Scence Review. 2003;**215**:265-273

[28] Meibom A, Yurimoto H, Cuif JP, Domart-Coulon I, Houlbrèque F, Constantz B, et al. Vital effect in coral skeletal composition display strict three-dimensional control. Geophysical Research Letters. 2006;**30**. DOI: 10.1029/2006GL025968

[29] Juillet-Leclerc A, Reynaud S, Rollion-Bard C, Cuif JP, Dauphin Y, Blamart D, et al. Oxygen isotopic signature of the skeletal microstructures in cultured corals: Identification of vital effects. Geochimica et Cosmochimica Acta. 2009;**73**:5320-5332

[30] Urey HC. Thermodynamic properties of isotopic substances. Journal of the Chemical Society. 1947:562-581

[31] Barnes DJ, Lough JM. Coral skeletons: Storage and recovery of environmental information. Global Change Biology. 1996;**2**:569-582

[32] Juillet-Leclerc A, Schmidt G. A calibration of the oxygen isotope paleothermometer of coral aragonite

from *Porites*. Geophysical Research Letters. 2001;**28**:4135-4138

[33] Kim ST, O'Neil JR, Hilaire-Marcel C, Mucci A. Oxygen isotope fractionation between synthetic aragonite and water. Influence of temperature and $Mg^{2+}$ concentration. Geochimica et Cosmochimica Acta. 2007;**71**:4704-4715

[34] Juillet-Leclerc A, Reynaud S. Light effects on the isotopic fractionation of skeletal oxygen and carbon in the cultured zooxanthellate coral, *Acropora*: Implications for coral-growth rates. Biogeosciences. 2010;7:893-906

[35] Marshall JF, McCulloch MT. An assessment of the Sr/Ca ratio on shallow water hermatypic corals as a proxy for sea surface temperature. Geochimica et Cosmochimica Acta. 2002;**66**:3263-3280

[36] D'Olivo JP, Sinclair DJ, Rankenburg K, McCulloch MT. A universam multi-trace element calibration for reconstructing sea surface temperatures from long-lived Porites corals: Removing "vital effects". Geochimica et Cosmochimica Acta. 2018;**239**:109-135

[37] Cohen AL, Layne GD, Hart SR, Lobel PS. Kinetic control of skeletal Sr/Ca in a symbiotic coral: Implications for the paleotemperature proxy. Paleoceanography. 2001;**16**:20-26

[38] Reynaud-Vaganay S, Gattuso JP, Cuif JP, Jaubert J, Juillet-Leclerc A. A novel culture technique for scleractinian corals: Application to investigate changes in skeletal $\delta^{18}O$ as a function of temperature. Marine Ecology Progress Series. 1999;**180**:121-130

[39] Gladfelter EH. Skeletal development in *Acropora cervicornis*: I. Patterns of calcium carbonate accretion in the axial corallite. Coral Reefs. 1982;**1**:45-51

[40] DeLong KL, Quinn TM, Taylor FW, Shen CC, Lin K. Improving coral-base paleoclimate reconstructions by replicating 350 years of coral Sr/Ca variations. Palaeogeography, Palaeoclimatology, Palaeoecology. 2013;**373**:6-24

[41] Quinn T, Sampson DE. A multiproxy approach to reconstructing sea surface conditions using coral skeleton geochemistry. Paleoceanography. 2002;**17**:1062. DOI: 10.1029/2000PA000528

[42] Stephans CL, Quinn TM, Taylor FW, Corrège T. Assessing the reproducibility of coral-based climate records. Geophysical Research Letters. 2004;**31**:L18210. DOI: 10.1029/2004GL020343

[43] Linsley BK, Messier RG, Dunbar RB. Assessing between-colony oxygen isotope variability in the coral *Porites lobata* at Clipperton Atoll. Coral Reefs. 1999;**18**:13-27

[44] Maier C, Felis T, Pätzold J, Bak RPM. Effect of skeletal growth and lack of species effects in the skeletal oxygen isotope climate signal within the coral genus *Porites*. Marine Geology. 2004;**207**:93-208

[45] Boiseau M, Juillet-Leclerc A. $H_2O_2$ treatment of recent coral aragonite: Oxygen and carbon isotopic implications. Chemical Geology. 1997;**143**:171-180

[46] Reynaud-Vaganay S, Juillet-Leclerc A, Gattuso JP, Jaubert J. Effect of light on skeletal $\delta^{13}C$ and $\delta^{18}O$ and interaction with photosynthesis, respiration and calcification in two zooxanthellate scleractinian corals. Palaeogeography, Palaeoclimatology, Palaeoecology. 2001;**175**:393-404

[47] Juillet-Leclerc A, Reynaud S. Light effects on the isotopic fractionation of skeletal oxygen and carbon in the cultured zooxanthellate coral, Acropora: implications for coral-growth rates. Biogeosciences. 2010;7:893-906

[48] Juillet-Leclerc A, Rollion-Bard C, Reynaud S, Ferrier-Pagès C. A new paradigm for δ18O in coral skeleton oxygen isotope fractionation response to biological kinetic effects. Chemical Geology. 2018;**483**:131-140

[49] Yu KF, Zhao JX, Wei GJ, Cheng XR, Chen TG, Felis T, et al. $\delta^{18}O$, Sr/Ca and Mg/Ca records of *Porites lutea* corals from Leizhou Peninsula, northern South China Sea, and their applicability as paleoclimatic indicators. Palaeogeography, Palaeoclimatology, Palaeoecology. 2005;**218**:57-73

[50] Corrège T. Sea surface temperature and salinity reconstruction from coral geochemical tracers. Palaeogeography, Palaeoclimatology, Palaeoecology. 2006;**232**:408-428

[51] Inoue M, Suzuki A, Nohara M, Hibino K, Kawahata H. Empirical assessment of coral Sr/Ca and Mg/Ca ratios as climate proxies using colonies grown at different temperatures. Geophysical Research Letters. 2007;**34**:L12611. DOI: 10.1029/2007GL029628

[52] Reynaud S, Ferrier-Pagès C, Meibom A, Mostefaoui S, Mortlock R, Fairbanks R, et al. Light and temperature effects on Sr/Ca and Mg/Ca ratios in the scleractinian coral *Acropora* sp. Geochimica et Cosmochimica Acta. 2007;**71**:354-362

[53] Suzuki A, Hibino K, Iwase A, Kawahata H. Intercolony variability of skeletal oxygen and carbon isotope signatures of cultured *Porites* corals: Temperature-controlled experiments. Geochimica et Cosmochimica Acta. 2005;**69**:4453-4462. DOI: 10.1016/j.gca.2005.05.018

[54] Armid A, Asami R, Fahmiati T, Sheikh MA, Fujimura H, Higuchi T, et al. Seawater temperature proxies based on DSr, DMg, and DU from culture experiments using the branching coral *Porites cylindrica*. Geochim Cosmochim Acta. 2011;**75**:4273-4285

[55] Adkins JF, Boyle EA, Curry WB, Lutringer A. Stable isotopes in deep-sea corals and a new mechanism for "vital effects". Geochimica et Cosmochimica Acta. 2003;**67**:1129-1143

[56] Gaetani GA, Cohen AL, Wang Z, Crusius J. Rayleigh-Based, Multi-Element Coral Thermometry: a Biomineralization Approach to Developing Climate Proxies. Geochim Cosmochim Acta. 2011;**75**:1920-1932

[57] Bertucci A, Moya A, Tambutté S, Allemand D, Supuran CT, Zoccola D. Carbonic anhydrases in anthozoan corals. A review. Bioorganic & Medicinal Chemistry. 2013;**21**:1437-1450

[58] Mass T, Drake JL, Peters EC, Jiang W, Falkowski PG. Immunolocalization of skeletal matrix proteins in tissue and mineral of the coral *Stylophora pistillata*. Proceedings of the National Academy of Sciences of the United States of America. 2014;**111**:12728-12733

[59] Allemand D, Tambutté E, Zoccola D, Tambutté S. Coral calcification, cells to reefs. In: Dubinski Z, Stambler N, editors. Coral Reefs: An Ecosystem in Transition. Springer Science + Business Media; 2011. pp. 119-150

[60] Uchikawa J, Zeebe RE. The effect of carbonic anhydrase on the kinetics and equilibrium of the oxygen isotope exchange in the $CO_2$-$H_2O$ system: Implications for $\delta^{18}O$ vital effects in biogenic carbonates. Geochimica et Cosmochimica Acta. 2012;**95**:15-34

[61] Ashman KM, Bird CM, Zepf SE. Detecting bimodality in astronomical datasets. Astronomy Journal. 1994;**108**. DOI: 10.1086/117248

[62] Moya A, Tambutté S, Tambutté E, Zoccola D, Caminiti N,

Allemand D. Study of calcification during a daily cycle of the coral *Stylophora pistillata*: Implications for light-enhanced calcification. The Journal of Experimental Biology. 2006;**209**:3413-3419

[63] Puverel S, Tambutté É, Pereira-Mouries L, Zoccola D, Allemand D, Tambutté S. Soluble organic matrix of two scleractinian corals: Partial and comparative analysis. Comparative Biochemistry and Physiology. 2005;**141B**:480-487

[64] Moya A, Tambutté S, Lotto S, Allemand D, Zoccola D. Carbonic anhydrase in the scleractinian coral *Stylophora pistillata*: Characterization, localization, and role in biomineralization. The Journal of Biological Chemistry. 2008;**283**:25475-25484

[65] McCrea JM. On the isotopic chemistry of carbonates and a paleotemperature scale. The Journal of Chemical Physics. 1950;**18**:849-857

[66] Al-Horani FA, Al-Moghrabi SM, de Beer D. The mechanism of calcification and its relation to photosynthesis and respiration in the scleractinian coral *Galaxea fascicularis*. Marine Biology. 2003;**142**:419-426

[67] Venn AA, Tambutté É, Holcomb M, Allemand D, Tambutté S. Live tissue imaging shows reef corals elevate pH under their calcifying tissue relative to seawater. PLoS One. 2011;**6**(5):e20013

[68] Mass T, Drake JL, Haramaty L, Kim JD, Zelzion E, Bhattacharya D, et al. Cloning and characterization of four novel coral acid-rich proteins that precipitate carbonates *In Vitro*. Current Biology. 2013;**23**:1126-1131

[69] Bertucci A, Forêt S, Ball EE, Miller DJ. Transcriptomic differences between day and night in *Acropora millepora* provide new insights into metabolite exchange and light-enhanced calcification in corals. Molecular Ecology. 2015;**24**:4489-4504

[70] Cai W, Borlace S, Lengaigne M, van Rensch P, Collins M, Vecchi G, et al. Increasing frequency of extreme El Niño eventsdue to greenhouse warming. Nature Climate Change. 2014;**4**:111-116

[71] Le Bec N, Juillet-Leclerc A, Correge T, Blamart D, Delcroix T. A coral $\delta^{18}O$ record of ENSO driven sea surface salinity in Fiji (south-western tropical Pacific). Geophysical Research Letters. 2000;**27**:3897-3900

[72] Shimamura M, Hyeong K, Yoo CM, Watanabe T, Irino T, Jung HS, et al. High resolution stable isotope records of sclreractinian corals near Ishigaki Island: Their implication as a potential paleoclimatic recorder in middle latitude regions. Geosciences Journal. 2008;**12**:25-31

[73] Cobb KM, Charles CD, Cheng HR, Edwards L. El Niño/southern oscillation and tropical Pacific climate during the last millennium. Nature. 2003;**424**:271-276